Introduction to Numerical Modeling in Earth and Planetary Sciences

Introduction to Numerical Modeling in Earth and Planetary Sciences

Christian Huber

Great Clarendon Street, Oxford, OX2 6DP,
United Kingdom

Oxford University Press is a department of the University of Oxford.
It furthers the University's objective of excellence in research, scholarship,
and education by publishing worldwide. Oxford is a registered trade mark of
Oxford University Press in the UK and in certain other countries

© Christian Huber 2025

The moral rights of the author have been asserted.

All rights reserved. No part of this publication may be reproduced, stored in a retrieval system, transmitted, used for text and data mining, or used for training artificial intelligence, in any form or by any means, without the prior permission in writing of Oxford University Press, or as expressly permitted by law, by licence or under terms agreed with the appropriate reprographics rights organization. Enquiries concerning reproduction outside the scope of the above should be sent to the Rights Department, Oxford University Press, at the address above.

You must not circulate this work in any other form
and you must impose this same condition on any acquirer.

Published in the United States of America by Oxford University Press
198 Madison Avenue, New York, NY 10016, United States of America

British Library Cataloguing in Publication Data

Data available

Library of Congress Control Number: 2025933836

ISBN 9780198802716
ISBN 9780198802723 (pbk.)

DOI: 10.1093/oso/9780198802716.001.0001

Printed by Printed and bound by
CPI Group (UK) Ltd, Croydon, CR0 4YY

The manufacturer's authorised representative in the EU for product safety is Oxford University Press España S.A. of El Parque Empresarial San Fernando de Henares, Avenida de Castilla, 2 – 28830 Madrid
(www.oup.es/en or product.safety@oup.com). OUP España S.A. also acts as importer into Spain of products made by the manufacturer.

Links to third party websites are provided by Oxford in good faith and for information only. Oxford disclaims any responsibility for the materials contained in any third party website referenced in this work.

Dedicated to Olga, Bénédicte, and Noémie.

Contents

Part III **Numerical modeling, partial differential equations (PDEs)**

Part IV **Overview of other numerical methods**

Part V **Acknowledgments**

List of Figures

List of Tables

Preface

Numerical modeling is becoming increasingly important in the physical sciences as we strive to quantify processes or develop mathematical models to explain data or observations. Although numerical modeling is attracting more attention in the Earth sciences, it remains a difficult topic to teach and it is sometimes felt as an overwhelming obstacle by students. The reasons behind this apprehension are that it blends not only Earth sciences, but also some aspects of coding, computer sciences, physics, and mathematics. This book is designed to provide a self-contained introduction to the art of numerical modeling in natural sciences, more specifically, in Earth and planetary sciences.

The emphasis of this book is not on coding or learning a coding language, although some of the pseudo-codes dispersed throughout the different chapters are written in a style that resembles Matlab, Python, or Julia, for simplicity. Rather, this book is designed to learn and demystify what I generally describe as the nine circles of Hell of modeling (borrowing an expression from Dante's *The Divine Comedy*), that is, the different steps that are required to start from the development of a mathematical model and finally reach the numerical solution of the model. Instead of the beautiful but dramatic illustrations that have been made for the path of Ovid and Dante to the Inferno, I decided, with help from my wife Olga, to opt for a more cheerful and perhaps modern (although a few decades old now) illustration of the challenges that heroes had to face back in the dark ages of video games (see Fig. 1). In that context each of the nine levels that faces our hero (you the reader) identifies a step or challenge that one faces when dealing with the derivation and implementation of a model.

The first step is the identification and formulation of the scientific question that drives the modeling effort. While this is the most exciting phase of research design for most—because it comes with the motivation to make progress on a question that we find worthy of effort—it is often challenging and generally comes more naturally with experience. The **second step** is more technical and involves the development of a mathematical description of the model proposed to address the question. This step generally involves differential equations, and analytical solutions are often beyond reach, which leaves numerical modeling as one of the logical alternatives. This book proposes and discusses several mathematical and physical models common to many scientific questions in Earth and planetary sciences. The **third step** is also technical and is given a thorough presentation in this book; it consists of transforming these differential equations into more computationally chewable algebraic sets of equations. This is done through various discretization techniques. While we use finite differences extensively throughout this book, other methods that rely on different discretization strategies are presented in Part IV. The **fourth step** is the search for an efficient and accurate way to solve the algebraic system of equations. This present choice of separating the selection of the discretization technique from the objective of finding a well-suited algorithm to solve for the variables of interest is arbitrary at best and, one may admit, partially motivated to count nine steps along our path as did Dante and Ovid (sorry about that). The **fifth step** is often considered by students taking this class as one of the two most infamous challenges in our journey: the implementation of boundary and initial conditions as required for the differential equations we aim to solve. The reason for this universal dread of boundary/initial conditions is that they are often related to the other infamous step (step 7). The **sixth step** is, with step 2, my personal favorite: it is now time to write your code and try to execute it. Why are we starting to code only now, six steps into the journey? I think it is important to avoid starting to write code before a suitable roadmap is designed—in other words, I seriously encourage you to work the kinks of your approach (steps 3–5) before coding becomes necessary. Then we arrive at the other most dreaded step, the **seventh step**: debugging. The chances you get your code working exactly as you hope on your first try are small: bugs come in different shapes and sizes, from a simple grammatical typo in your code to a conceptual error. The art of debugging takes time to master and it is based on experience. Strangely, there may be something hopeful there: the more bugs you tend to slide into your codes the faster you will master debugging. Once the code seems to work well, you need to remove any remaining doubt of its performance by validating it with analytical solutions (the **eighth step**). This is not easy, but it is fundamentally important. A model is only good to use if it is properly validated. The issue though is that if we write a code to start with it is because the problem we want to solve does not admit an analytical solution. Otherwise, why should we bother? An ingenious way to overcome this issue is to design a suite of simpler experiments that test each aspect of your model and where analytical solutions exist. This is often the best way to go. Once that last hurdle is cleared you have reached the ultimate goal, on the **ninth step**, where it is time to enjoy addressing the scientific question that has motivated this ordeal.

These different steps are necessary to understand and be in control of the modeling effort you are about to undertake. I believe they provide a blueprint for a rigorous path to successful modeling. This book is intended to guide you through these steps while applying modeling to simple problems relevant for Earth and planetary sciences. As mathematics is the language of physical modeling, a whole section of this book is devoted to the mathematical tools required for our journey. This section should not serve as a substitute for a more in-depth mathematical reference, but should provide enough knowledge and intuition to go through the book and build a sound logical understanding of modeling.

The material for this book was developed over a decade of teaching numerical modeling to Earth, Planetary and Environmental Scientists and Engineers at Georgia Tech and then Brown University. The content of this book has been significantly improved by comments, questions, and suggestions from students and colleagues, as well as the editing staff at Oxford University Press. In a book like this one, any mistakes, imprecisions, and vagueness are unfortunately present, and I take sole responsibility for these.

The material for this book has been used for a semester-long course at the level of advanced undergraduate and graduate students. The mathematical section is mostly used to level potential gaps in knowledge that arise as we move forward and I suggest using that section as support material, rather than the focus of lectures. In the semester format for this course, the treatment of ordinary differential equations (ODEs) takes the first half, while the second half is focused on the solution of partial differential equations (PDEs). My experience is that a term project where the concepts introduced are applied to a problem of personal interest for the student is an optimal way to complement the lectures.

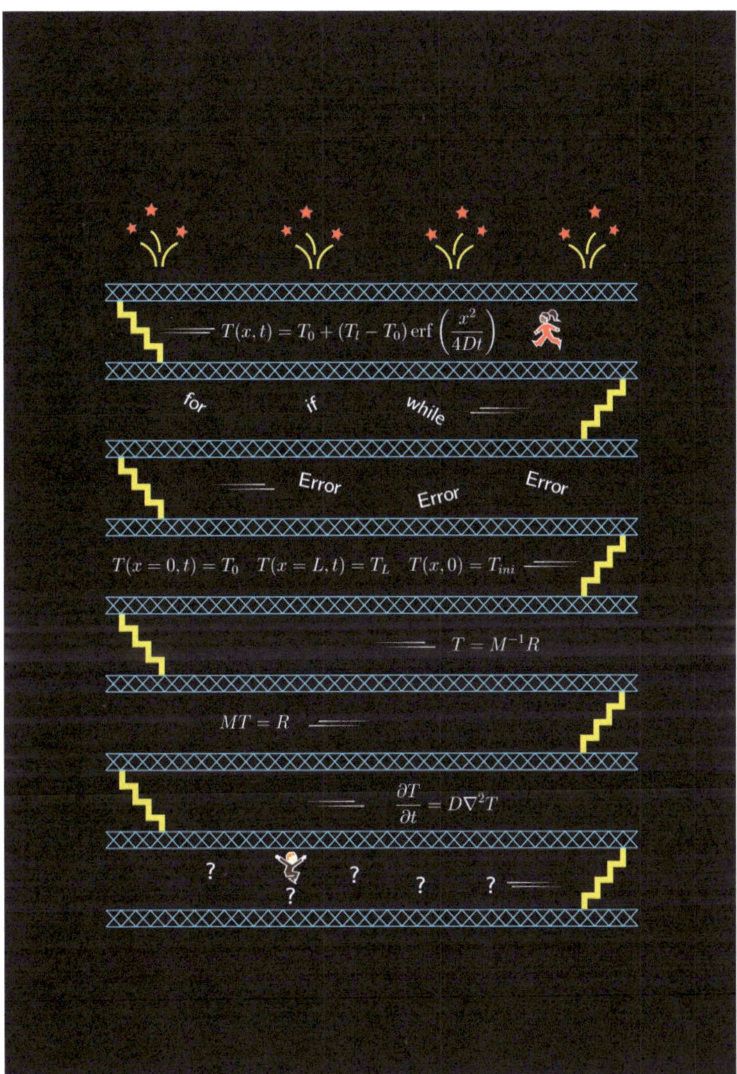

Figure 1 A "modern" version of Dante's nine circles of Hell applied to the principles of numerical modeling. From problem conceptualization (circle-level 1) to solution (circle-level 9).

Part I

Mathematical concepts

Introduction to real valued calculus

0.1 Continuous functions

In mathematics a function is an object that transforms an element (a number or a vector) of a set to an element of another set. This transformation is unique in the sense that each element of the originating set is linked to a single element of the destination set. This is a rather abstract way to start, and it is mathematically codified assuming the function $f(x)$ is defined for real numbers $x \in \mathbb{R}$:

$$f : \mathbb{R} \; to \; \mathbb{R}, \quad x \mapsto f(x). \tag{0.1}$$

In a more intuitive (but perhaps less exact) way, a function is a mapping from a number to another number or from a set of numbers (vector) to another set of numbers. We are quite familiar with some of these functions, for instance, trigonometric functions cos, sin, tan, or polynomial functions such as $f_1(x) = x^2 + 3x$. Some of these transformations are linear, that is, if applied to two elements x_1 and x_2 such that $x_2 = \alpha x_1$, then $f_2(x_2) = \alpha f(x_1)$, while this would not be the case for f_1 above (not a linear function).

Another fundamental property of functions is whether or not they are continuous. The common graphic illustration of continuous functions is one where it is possible to draw the function without lifting the pencil from the page (i.e., no jumps; see both functions on left panel of Figure 0.1). A more mathematical definition for continuity is drawn from the concept of mathematical limits. In the context of functions the limit of the function $f(x)$ when x approaches a value x_0 is the value the function approaches as x is brought closer and closer to the value x_0, it is generally written $\lim_{x \longrightarrow x_0} f(x)$. Continuous functions are functions that satisfy

$$\lim_{x \longrightarrow x_0} f(x) = f(x_0), \tag{0.2}$$

for any x_0 that belongs to the domain of the function (e.g., real numbers). Even if this more exact definition seems more complicated, it aligns with the more intuitive illustration mentioned above, as we draw the function and approach the value x_0, then the function $f(x)$ is not allowed to jump and it gradually reaches the value $f(x_0)$.

Why do we care about continuous functions for numerical modeling? We can define the slope of a continuous function at any point x within its domain, and that, as we see later, allows us to write the rate of change of this function (i.e., its slope) as derivatives. This may sound a bit trivial or even irrelevant, but actually it is a property that we take advantage of routinely in mathematical modeling. Mathematical models in physical sciences are rooted in the expression of differential equations (i.e., equations with derivatives of continuous functions), as we seek to understand how observables vary in time and space in any problem of interest. Therefore, if differential calculus is the language of physical sciences, then differential equations are its sentences and derivatives of functions are the words that provide meaning to this language. Derivatives are only well defined if the differentiated functions are continuous.

Introduction to Numerical Modeling in Earth and Planetary Sciences, Christian Huber, Oxford University Press.
© Christian Huber (2025). DOI: 10.1093/oso/9780198802716.003.0001

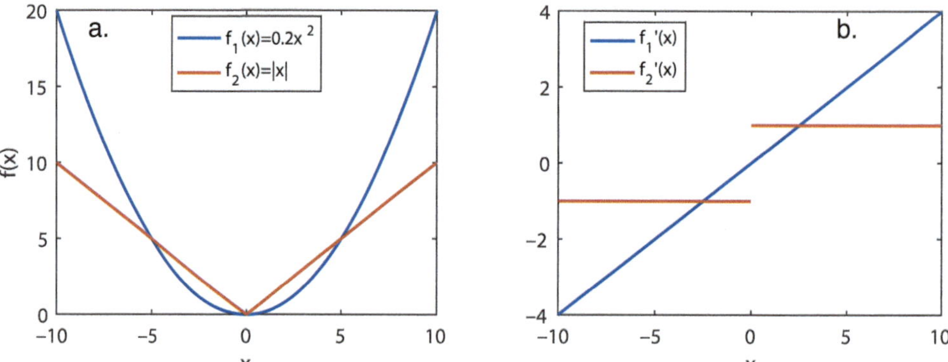

Figure 0.1 Example of continuous functions (left panel) and their derivative (right panel). Functions f_1 and f_2 are both continuous and while derivative f_1 derivative is continuous, derivative f_2 is not.

0.1.1 **Exercises**

- Show the continuity of the function $f(x) = x^2 + 3x$ at $x = 2$. For that, consider the difference between the limits of $x \to 2^+$ (where $x > 2$ and is decreasing towards 2) and $x \to 2^-$ (where $x < 2$ and is increasing towards 2).
- Show the continuity of the absolute value function $g(x) = |x|$ at $x = 0$.
- Discuss the continuity (or lack thereof) of the slope of $g(x)$ in the exercise just above. Where is the slope of $g(x)$ not continuous?

0.2 **Roots of a function**

Another important consideration about a function is whether it admits elements x for which the function is null $f(x) = 0$. These elements x are called roots of the function and nonlinear functions can have more than one root. Alternatively, nonlinear functions can also have no roots, for example, the exponential function $f(x) = \exp(x)$ only approaches 0 when $x \to -\infty$. Finding the root(s) of a function can be tricky mathematically and finding general solutions to find roots of nonlinear functions has been historically one of the main challenges of mathematicians of only a few centuries ago. In high school we learn how to find the roots of quadratic functions such as $f(x) = ax^2 + bx + c$ by applying the approach of the discriminant Δ, where $\Delta = b^2 - 4ac$ and the roots are

$$x_1 = \frac{-b + \sqrt{\Delta}}{2a} \tag{0.3}$$

$$x_2 = \frac{-b - \sqrt{\Delta}}{2a}. \tag{0.4}$$

Note that these roots can be real or complex numbers. Things get more complicated with nonlinear functions that are not quadratic. Numerical approximations are often used to retrieve roots of functions and we discuss some approaches later on.

0.3 Derivatives, local and global extrema

0.3.1 Derivative of a function, definition

The derivative of a single variable function is its slope at a given point. Because the derivative of a continuous function is itself a function (i.e., it maps a slope value to any point along that continuous function), the concepts of limits and roots described above apply as well. The average slope of a function $f(x)$ over the interval $[x, x + \delta x]$ is defined as the change in the function over that interval divided by the interval itself. For example, if the function $f(x)$ describes the elevation of a one-dimensional terrain, then the average slope is the change in elevation divided by the horizontal distance covered. Mathematically this amounts to

$$\text{average slope} = \frac{f(x+\delta x) - f(x)}{(x+\delta x) - x} = \frac{f(x+\delta x) - f(x)}{\delta x}. \tag{0.5}$$

The derivative is the limit of the slope a one point (here x), that is, when the interval $\delta x \to 0$. Mathematically that amounts to defining the derivative of the function $f(x)$ at x as

$$\frac{df}{dx}(x) = \lim_{\delta x \to 0} \frac{f(x+\delta x) - f(x)}{\delta x}. \tag{0.6}$$

Through this definition it is easy to retrieve some of the well-known identities of derivatives of commonly used functions, for example, $f(x) = x^n$. Indeed,

$$\frac{dx^n}{dx}(x) = \lim_{\delta x \to 0} \frac{(x+\delta x)^n - x^n}{\delta x}, \tag{0.7}$$

using $(x + \delta x)^n = x^n + n\delta x x^{n-1} + \frac{1}{2}n(n-1)\delta x^2 x^{n-2} + \ldots + \delta x^n$, we can reorganize terms and cancel the contribution in x^n to get

$$\frac{dx^n}{dx}(x) = \lim_{\delta x \to 0} \frac{\delta x \left(nx^{n-1} + \frac{n(n-1)}{2}\delta x x^{n-2} + \cdots + \delta x^{n-1} \right)}{\delta x}. \tag{0.8}$$

The extended sum is made of terms that are at least of order δx so when simplifying the δx on the numerator and denominator we retrieve

$$\frac{dx^n}{dx}(x) = \lim_{\delta x \to 0} \left(nx^{n-1} + \frac{n(n-1)}{2}\delta x x^{n-2} + \cdots + \delta x^{n-1} \right) = nx^{n-1} \tag{0.9}$$

where all the terms except the first one vanish because they depend on δx.

Another example here is the derivative of $\cos x$, which (according to the definition of derivatives) is

$$\frac{d\cos x}{dx}(x) = \lim_{\delta x \to 0} \frac{\cos(x+\delta x) - \cos(x)}{\delta x}. \tag{0.10}$$

Using the trigonometric identity $\cos(\alpha + \beta) = \cos\alpha\cos\beta - \sin\alpha\sin\beta$, we can recast the first term on the numerator and get

$$\frac{d\cos x}{dx}(x) = \lim_{\delta x \to 0} \frac{\cos x \cos\delta x - \sin x \sin\delta x - \cos x}{\delta x} \tag{0.11}$$

Grouping the terms in $\cos x$ we get

$$\frac{d\cos x}{dx}(x) = \lim_{\delta x\to 0} \frac{\cos x(\cos\delta x - 1) - \sin x \sin\delta x}{\delta x}. \quad (0.12)$$

This can be recast as

$$\frac{d\cos x}{dx}(x) = \lim_{\delta x\to 0} \frac{\cos x(\cos\delta x - 1)}{\delta x} - \sin x\frac{\sin\delta x}{\delta x} \quad (0.13)$$

where the first fraction can easily be shown to become vanishingly small in the limit $\delta x \to 0$ and using the fact that

$$\lim_{\delta x\to 0} \frac{\sin\delta x}{\delta x} = 1, \quad (0.14)$$

leads to the well-known identity

$$\frac{d\cos x}{dx}(x) = -\sin x. \quad (0.15)$$

The derivative of a function is a function itself and it may or may not be continuous (see an example of the latter with the derivative of the absolute value function). When the derivative is continuous it can be differentiated again to obtain a second-order derivative and the process can be repeated as long as the resulting function is continuous (yielding higher-order derivatives). For most cases of interest in continuum mechanics, on which most examples in this book rely, functions and their derivatives are continuous, which allows us to differentiate them as needed.

0.3.2 **Extrema and roots of derivatives**

It is often useful to study the existence and calculate the position of extrema in continuous functions, that is, locations where they reach a local or global maximum/minimum. For example, minima in Gibbs free energy in thermodynamics define equilibrium states. Extrema of a function $f(x)$ can be easily identified as the roots of the derivative function $df/dx(x)$. Indeed, a minimum is defined by slopes that become increasingly more negative to the left and more positive to the right of the minimum. Given the derivative function is continuous and the slope is the local value of the derivative function, we infer that the derivative function is negative before a minimum and positive after and therefore has to go through zero at the minimum if it is continuous. The same argument can be made for a maximum, with the slopes to the left being positive and negative to the right (curvature is the opposite of a minimum).

Extrema of functions are roots of their derivative function. Is it then appropriate to expect all roots of the derivative function to be an extrema? Actually it is not, as a saddle point, that is, a location in a function where it changes curvature sign (e.g., goes from concave upward to concave downward) would also yield a root of the derivative function. An example is given by the function $f(x) = ax^3$, where a is any nonzero constant. Its derivative is given (see above) by $df/dx(x) = 3ax^2$, which admits a root at $x = 0$. Upon inspection of Figure 0.2 we can see that $f(x = 0)$ is not an extrema of $f(x)$, but rather a saddle point. So how do we know if $df/dx(x) = 0$ corresponds to a minimum, a maximum or a saddle point? There are several options. One is to graph the function and interpret it as we did in Figure 0.2. A perhaps more useful approach is to consider the derivative of the derivative function (i.e., the second derivative of the original function $f(x)$).

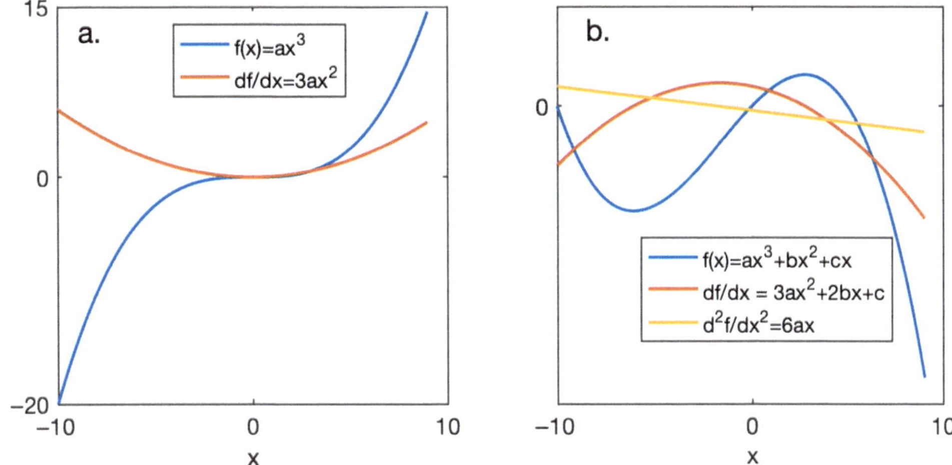

Figure 0.2 a): Example of a function $f(x)$ with a saddle point at the origin. The derivative of $f(x)$ also shown in the figure has a root at the origin. b): Example of a function with a local minimum and local maximum. The local minimum is located to the left of the origin and the maximum on its right. Interestingly the sign of the second derivative of $f(x)$, d^2f/dx^2 changes signs (from positive to negative) at the origin.

As illustrated by an example on Figure 0.2b, the function $f(x) = ax^3 + bx^2 + cx$ has three roots: one at the origin $x_1 = 0$ and two roots at $x_2 = (-b - \sqrt{b^2 - 4ac})/2a$ and $x_3 = x_2 = (-b + \sqrt{b^2 - 4ac})/2a$. Its derivative function df/dx has two roots: one at $x_4 = (-b - \sqrt{b^2 - 3ac})/3a$ and the other at $x_5 = (-b + \sqrt{b^2 - 3ac})/3a$. This example illustrates that minima of $f(x)$ are found when the derivative function df/dx admits a root and the second derivative function d^2f/dx^2 is positive (i.e., positive curvature), while maxima are obtained at locations that are roots of the derivative function df/dx and negative values for the second derivative function d^2f/dx^2. Saddle points exist at locations where the derivative and the second derivative functions are both nil. As a general rule we can use the sign of the second derivative to sort out the type of extrema or to distinguish them from saddle points.

While differentiating a continuous function is rather straightforward, finding the roots of functions can be tricky. Numerical approximations are often used to calculate these points within a tolerance. There are many possible paths to obtain these numerical approximations, and they are all based on iterative procedures where an initial guess x_0 defines the starting point and the root's approximated position x_k is refined as many times (here k iterations) until the function's evaluation at the approximated root value is small enough, that is,

$$|f(x_k)| < Tol, \tag{0.16}$$

where Tol is the preset tolerance (accuracy of the approximation in terms of the value of the function). This does not mean that we are necessarily very close to the desired solution (here the root), because for relatively flat functions (i.e., functions where the slope around the root is very low), a small error on the value of $f(x)$ does not translate in a small departure from the position of the root.

Using the definition of the average slope between two points

$$slope = \frac{f(x_a) - f(x_b)}{x_a - x_b}, \tag{0.17}$$

if now x_b is a root of $f(x)$, then

$$slope = \frac{f(x_a)}{x_a - x_b}, \tag{0.18}$$

and manipulating the equation we get

$$f(x_a) = slope(x_a - x_b). \tag{0.19}$$

This illustrates the point made here, because for $f(x_a)$ to approach 0 it does not necessarily require x_a to approach x_b very closely. Actually, if the slope is very small around the root there could be a significant gap between the two positions (although this depends on the tolerance). In essence, this simple argument highlights that both the slope and the value of $f(x)$ around the root are important constraints in finding a suitable approximation for the root of a function. This provides a natural transition to the Newton–Raphson iterative method, which is perhaps the most commonly used root-finding numerical approximation. Starting from a conceptual illustration of the method helps. Let us consider the two different cases presented in Figure 0.3 (panels a and b), with two different initial guesses x_0 and x_0*. The functions $f(x)$ and $g(x)$ are nonlinear (sin curves) and have generally different slope signs around their root.

Let us start from an initial guess for the position of the root of $f(x)$, which we define as x_0 (see Figure 0.3a). Upon rapid inspection (either by computing the value $f(x_0)$ or graphically), we see that $f(x_0)$ is negative and not particularly close to 0. The iteration procedure is to start from a guess

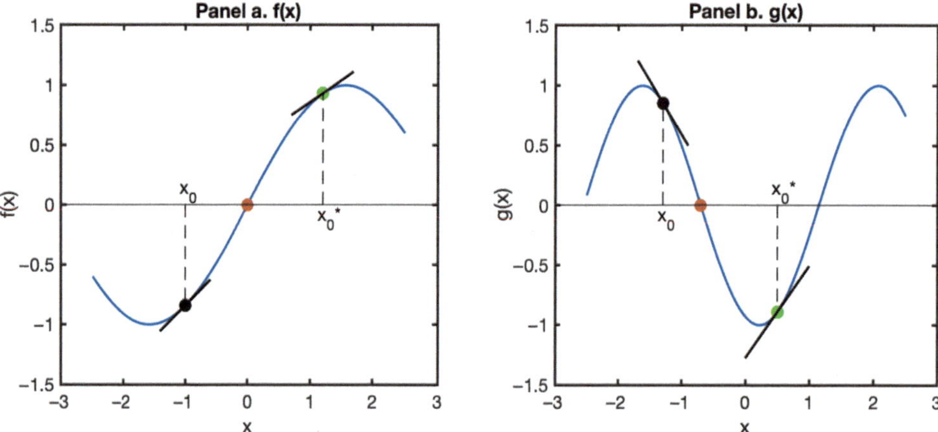

Figure 0.3 Finding roots with Newton–Raphson method. A first guess at x_0 is updated iteratively to get closer to a local root of the function.

x_0 and compute a hopefully improved guess by

$$x_1 = x_0 + \Delta x_{0-1}, \tag{0.20}$$

where Δx_{0-1} is a perturbation to move along the x-axis and hopefully get closer to the root of $f(x)$. For the moment we assume that there is only one root to be found (we revisit this assumption later in the chapter). How do we define the perturbation to build a new iteration of the guess? Starting from the definition of the average slope of a function between two points (here, x_0 and the target x_1),

$$slope_{0-1} = \frac{f(x_1) - f(x_0)}{x_1 - x_0}, \tag{0.21}$$

and, being greedy, let us assume for an instant that x_1 is the root we seek, then reorganizing the Eq. 0.21 and using $f(x_1) = 0$ leads to

$$x_1 = x_0 - \frac{f(x_0)}{slope_{0-1}}, \tag{0.22}$$

and that means that

$$\Delta x_{0-1} = -\frac{f(x_0)}{slope_{0-1}}. \tag{0.23}$$

There is a catch, though. We cannot compute the slope between x_0 and x_1 because we are trying to figure out where x_1 is along the x-axis. So instead of a single step to get to the root, we can approximate the slope of the function $f(x)$ between x_1 and x_0 with a slope we can compute from the definition of the function $f(x)$ alone. If the function $f(x)$ is defined mathematically, we can readily compute the local slope of $f(x)$ at x_0 our initial guess, which is given by the derivative of $f(x)$ at x_0. Obviously this is only an approximation of the average slope between x_1 and x_0, and if the function is continuous (as it should be!), then the approximation gets better with the distance between x_1 and x_0 being smaller. The cost of this assumption is that, for most functions, x_1 will no longer be the root and that, with this procedure, we have just moved (hopefully) closer to the root. Mathematically, the procedure is to replace the definition of Δx_{0-1} by the more tractable

$$\Delta x_{0-1} = -\frac{f(x_0)}{df/dx(x_0)}. \tag{0.24}$$

Before moving on to establish the iterative nature of this approximation, we must visualize with examples what this correction to our first guess actually achieves. Going back to Figure 0.3, and starting with panel (a) and the function $f(x)$, if the initial guess is x_0, we see that $f(x_0) < 0$ and the slope $df/dx(x_0)$ at x_0 is positive. By inspection here, we would like $\Delta x_{0-1} > 0$ to move closer to the root (at least in the right direction). We see that this is accomplished by the definition of Eq. 0.24. Now moving to the other guess x_0*, which is located to the right of the root, we see that $f(x_0*) > 0$ and the slope is also positive at this point. In that case, $\Delta x_{0-1} < 0$, which leads to $x_1 < x_0*$ and the iteration moves us to the left along the x-axis. This is reassuring and the same can be done with the guess x_0 and the function $g(x)$ on Figure 0.3b. So this simple procedure moves the search of the root in the correct direction. Moreover, the magnitude of the step Δx_{0-1} is controlled by the magnitude of $f(x_0)$ and the slope $df/dx(x_0)$. Considering first $f(x_0)$, if its magnitude is large, we can expect x_0 to be relatively far away from the root (again, the function is continuous) and would therefore prefer a relatively larger step to move closer to it. Conversely, if $f(x_0)$ is small then we may not want a large step away from x_0 because x_0 may be close to the root. The dependence of $\Delta x_{0\text{-}1}$ on $f(x_0)$ is therefore sound. Considering now the effect of the slope of the function at x_0, if $f(x_0) \neq 0$ and the slope $df/dx(x_0)$ is large in magnitude, a small step $\Delta x_{0\text{-}1}$ may be warranted to avoid overshooting the root. But if the slope is small in magnitude, a large step may be preferable because we need to move further away from x_0 to get close enough to the root. Again, the choice of $\Delta x_{0\text{-}1}$ makes intuitive sense.

Now that we are confident that the procedural step $x_1 = x_0 - \frac{f(x_0)}{df/dx(x_0)}$ is sound, we can build an iterative procedure:

$$
\begin{aligned}
& x_0 \\
& x_1 = x_0 - \frac{f(x_0)}{df/dx(x_0)} \\
& x_2 = x_1 - \frac{f(x_1)}{df/dx(x_1)} \\
& \dots \\
& x_k = x_{k-1} - \frac{f(x_{k-1})}{df/dx(x_{k-1})},
\end{aligned}
\tag{0.25}
$$

which we repeat until $|f(x_k)|$ becomes smaller than a preset tolerance. This is called the Newton–Raphson method, and the convergence rate of this iterative algorithm (i.e., how many steps k are required to get a satisfying result) depends largely on the function being studied and the initial guess.

Here is a pseudo-code to implement the Newton–Raphson root finding method:

```
x0=1; % Initial guess
Tol=1e-6;                  % Tolerance
Error_f=2*Tol;         % Initialization of error
x_old=x0; % initialization for 1st step
while Error_f > Tol  % loop for NR iteration
  x_new=x_old-f(x_old)/df(x_old); % df is df/dx
  Error_f = f(x_new);
  x_old=x_new;
end
```

There are a few caveats though, and it can be seen in Figure 0.3b with the initial guess x_0*. Computing the step $\Delta x_{0\text{-}1}$ in this case gives a positive step value and moves the approximation away from the root highlighted by the dot. When dealing with functions with multiple roots or with multiple local extrema, the procedure here can get stuck in local or global minima away from the desired root. A second possible issue is if any guess x_k is such that $df/dx(x_k) = 0$, the step there becomes infinite and this is the end of the game. It is therefore essential to be careful with this procedure.

0.3.3 **Exercises**

- Find the all roots of $f(x) = 3x^3 - 10x^2 + 6x$ analytically. How many extrema does $f(x)$ admit and where are they located? Analyze the nature of the extrema both mathematically (using the second derivative of $f(x)$) as well as graphically.
- Implement a Newton–Raphson script to approximate the root of $f(x)$ defined in the previous exercise. Try to retrieve the roots derived above by playing with the position of your initial guess.
- Run the Newton–Raphson script with $f(x)$ defined here varying the initial guess between $-1 \leq x_0 \leq 3$ by increments of 0.1. Plot and compare (a) the function $f(x)$ over this interval and (b) the converged root value obtained from the Newton–Raphson solver as function of the initial guess x_0. Discuss what you see.

0.4 **Derivatives and integrals: The fundamental theorem of calculus**

This section establishes a formal link between derivatives and integrals, which may be dismissed as a trivial identity but rather is the basis of the fundamental theorem of calculus (FTC); further, it is a much deeper concept than often acknowledged. It is also deeply relevant to numerical models because the basic language of continuum mechanics is formalized by differential equations and our solution procedure is to integrate these equations to retrieve solutions. This is in essence possible *because of* the FTC.

The idea behind the FTC is that if the derivative of the function is related to the tangent (slope) of a function $f(x)$ at any position x, then integrals are the opposite operation. Interestingly, while derivatives are "easy" to compute on most analytical continuous functions, integrals are often difficult to calculate (we return to this point later). The FTC provides a simple alternative, by guessing the proper primitive function.

The antiderivative $F(x)$ (or primitive) of a function $f(x)$ is defined by

$$\frac{dF}{dx}(x) = f(x). \tag{0.26}$$

We see later that there is an infinite choice of primitive $F(x)$ to any proper function $f(x)$. One of the statements of the FTC is that there is an operator "$\int$" that can act on the function $f(x)$ such that

$$\int_a^b f(x)\mathrm{dx} = f(b) - f(a). \tag{0.27}$$

An illustration is needed to build intuition into this expression. Let us imagine that we are riding a bus from an initial bus stop at position x_a to another bus top at x_b. For simplicity, we assume that the bus travels in a pseudo one-dimensional world. The instantaneous velocity of the bus $V(t)$ is defined over the time interval $[t_a, t_b]$ where $x(t_a) = x_a$ and $x(t_b) = x_b$. If we log the instantaneous velocity $V(t)$ over time during the transport, a plausible graphic representation of this record is provided in Figure 0.4. If the velocity logs are regular entries separated by a uniform time interval

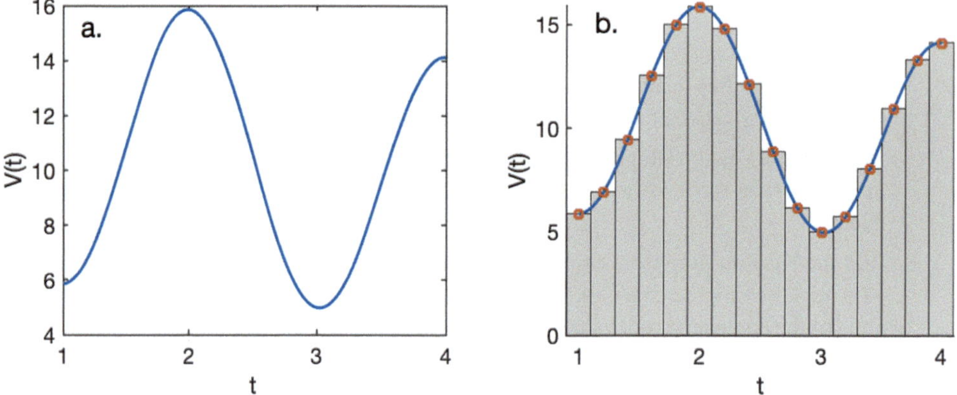

Figure 0.4 a): Velocity of the bus over time. b): Velocity averaged over small increments of time (panel b). Images show the relation of the derivative of the position with time (velocity) with the integral of the velocity.

$\delta t << t_b - t_a$, we can compute the average velocity over the travel

$$\langle V \rangle = \frac{1}{t_b - t_a} \sum_{k}^{N} V(t_k)\delta t, \tag{0.28}$$

where $t_k = t_a + (k + 1/2)\delta t$ and the sum is taken over the number of intervals δt in the time interval $t_a - t_b$. Because we are interested in the instantaneous velocity $V(t)$ and want to compute the most accurate average velocity, it becomes advantageous to reduce $\delta t \to 0$, bringing the concept of limits. Note that when δt decreases the number of intervals required to span $t_a - t_b$ increases proportionally. We define the operator integral as the limit of the sum under these conditions

$$\int_{t_a}^{t_b} V(t)\mathrm{d}t = \lim_{\delta t \to 0} \sum_{k} V(t_k)\delta t. \tag{0.29}$$

With this formalism the average velocity can be related to the integral

$$\langle V \rangle = \frac{1}{t_b - t_a} \int_{t_a}^{t_b} V(t)\mathrm{d}t. \tag{0.30}$$

Alternatively, if the distance between the two bus stops ($\Delta x = x_b - x_a$) and the duration of the travel ($\Delta t = t_b - t_a$) are known, then the average velocity can be retrieved from

$$\langle V \rangle = \frac{\Delta x}{\Delta t} = \frac{x_b - x_a}{t_b - t_a}. \tag{0.31}$$

Equating both expression leads to

$$\frac{x_b - x_a}{t_b - t_a} = \frac{1}{t_b - t_a} \int_{t_a}^{t_b} V(t)\mathrm{d}t, \tag{0.32}$$

which ultimately gives us

$$\int_{t_a}^{t_b} V(t)\mathrm{dt} = x(t_b) - x(t_a), \tag{0.33}$$

which is one expression of the FTC. This means that the position $x(t)$ is an antiderivative of the instantaneous velocity $V(t)$ with respect to time and therefore it is no surprise that

$$\frac{dx}{dt} = V(t). \tag{0.34}$$

By extension and using Eq. 0.33 we can express the antiderivative at x, $F(x)$ as

$$F(x) - F(a) = \int_a^x f(x')\mathrm{dx}', \tag{0.35}$$

which we can differentiate with respect to x to get

$$\frac{dF}{dx} = \frac{d}{dx}\left[\int_a^x f(x')\mathrm{dx}'\right]. \tag{0.36}$$

Using the definition of the antiderivative $f(x)$, we get

$$\frac{d}{dx}\left[\int_a^x f(x')\mathrm{dx}'\right] = f(x). \tag{0.37}$$

This demonstrates that derivative and integrals are opposite operators. Moreover, as mentioned, integrals are more cumbersome to compute than derivatives and this relation between the two operators offers an easier way to compute integrals, as they can be estimated by guessing proper antiderivative functions. For example, to a constant, the relation

$$\int 3x^2\mathrm{dx} = x^3 \tag{0.38}$$

can be computed by remembering that the derivative of x^n is nx^{n-1} with $n = 3$ here. Another important result of the FTC is that any function that can be written as the derivative of an antiderivative (or primitive) function

$$\frac{dF}{dx}(x) = f(x), \tag{0.39}$$

is such that its definite integral

$$\int_a^b f(x)\mathrm{dx} = F(b) - F(a), \tag{0.40}$$

which introduces somehow a notion of path invariance (which is central to the idea of conservative forces in physics). For instance, electric and gravitational forces are conservative forces because they can be expressed as the gradient of a potential (electric potential or gravitational potential, respectively). This means that if we conduct an integral over a two- or three-dimensional path that forms a closed loop, these forces do not exert a net work

$$\oint_{\vec{x}_a} \vec{F}(\vec{x}) \cdot \mathrm{d}\vec{l} = U(\vec{x}_a) - U(\vec{x}_a) = 0, \tag{0.41}$$

where $\oint$ is the sign for a closed integral in two or three dimensions, $\vec{F}$ is the force field, $\mathrm{d}\vec{l}$ is an element of the closed path, and $U(\vec{x})$ is the potential (anti-derivative) that relates to the force $\vec{F}$. This would not be true for a friction force, for example, $\vec{F} = \mu\vec{V}$, where μ is a friction constant and $\vec{V}$ the velocity field of the object. This force cannot be expressed as the total derivative of a potential function (antiderivative) and therefore will exert a net energy loss even for closed paths.

The last point worth commenting relates back to the continuity of the function $f(x)$. We saw earlier that for a derivative function $f'(x) = df/dx(x)$ to exist, the function $f(x)$ needs to be continuous at all points in the domain where it is defined. That condition is not a requirement to compute the integral of $f(x)$. Indeed the integral of $f(x)$ can be computed and yields a continuous function even if $f(x)$ is discontinuous at some points within the interval. We briefly discuss this in the last part of this book while presenting a brief overview of other methods that take advantage of the weak form, that is, integrate spatially the differential equation to smooth the solution procedure.

0.5 Approximation of continuous functions: Taylor series

This section introduces a key mathematical concept in numerical modeling: the approximation of continuous functions with Taylor series. We defined the derivative of a function and showed earlier that, if the derivative is itself a continuous function, then it can be differentiated again (i.e., second derivative of the original function, see Section 0.3.2). The procedure can be pursued further as long as the differentiated function remains continuous. Imagine that you are moving along a one-dimensional landscape and have knowledge about the local topography, the elevation, slope, and change in slope (second derivative) only at the location you are standing (no peeking to see what is ahead of you!). You can probably make an educated guess as to the terrain very near your location, for example, is the elevation likely to be increasing or decreasing? Using logic and your knowledge of the function at your location (especially the slope of the elevation), you can assume for example that the elevation is likely to be higher if the derivative along the direction you plan to travel is positive and lower if it is negative. That may not be exact, but it is a decent starting point.

The idea of the Taylor series is to define a continuous function $f(x)$ at any position x from knowledge of the value of the function and its derivatives at a reference position x_0. For example, to a zeroth order we can assume the function $f(x)$ to be incredibly dull and constant for all x, so that

$$f(x) = f(x_0). \tag{0.42}$$

This obviously is not correct for most functions. Using the argument laid out earlier, we could use the local slope at x_0 to infer a likely correction for $f(x)$, that is, there is a finite slope, and it is possible the value of $f(x) \neq f(x_0)$ and get a first-order correction

$$f(x) \approx f(x_0) + \frac{df}{dx}(x_0)(x - x_0), \tag{0.43}$$

where the correction is the slope at x_0 times the displacement along the axis, which gives us $\Delta f = f(x) - f(x_0)$. This first-order estimate is illustrated in Figure 0.5 using a sine curve as an example and $x_0 = 2$.

Correcting the value of $f(x)$ by the slope assumes that the "elevation" ($f(x)$) does not have any curvature around the region of interest. If the slope changes (curvature) around x_0 then we may expect that accounting for it would improve our estimation of $f(x)$, and therefore we can add a new (second-order) correction

$$f(x) = f(x_0) + \frac{df}{dx}(x_0)(x - x_0) + \frac{d^2f}{dx^2}(x_0)\frac{(x - x_0)^2}{2}. \tag{0.44}$$

The correction now is shown in Figure 0.6. This process can be repeated as long as derivatives of derivatives remain continuous functions (see Figure 0.7) and provides more accurate approximation of $f(x)$ even further away from the reference position x_0.

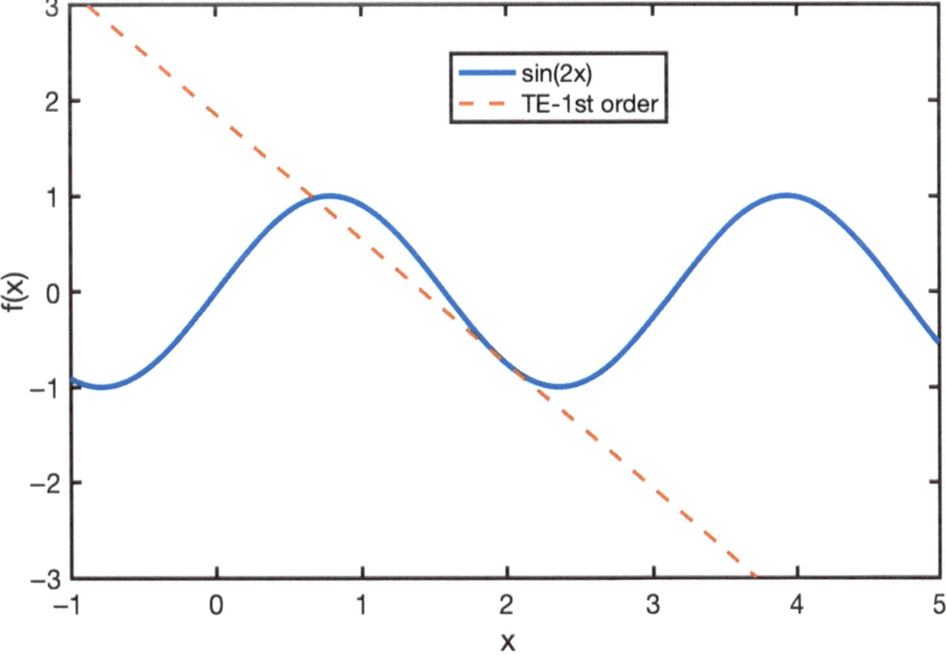

Figure 0.5 First-order Taylor approximation of $\sin(2x)$ around $x = 2$.

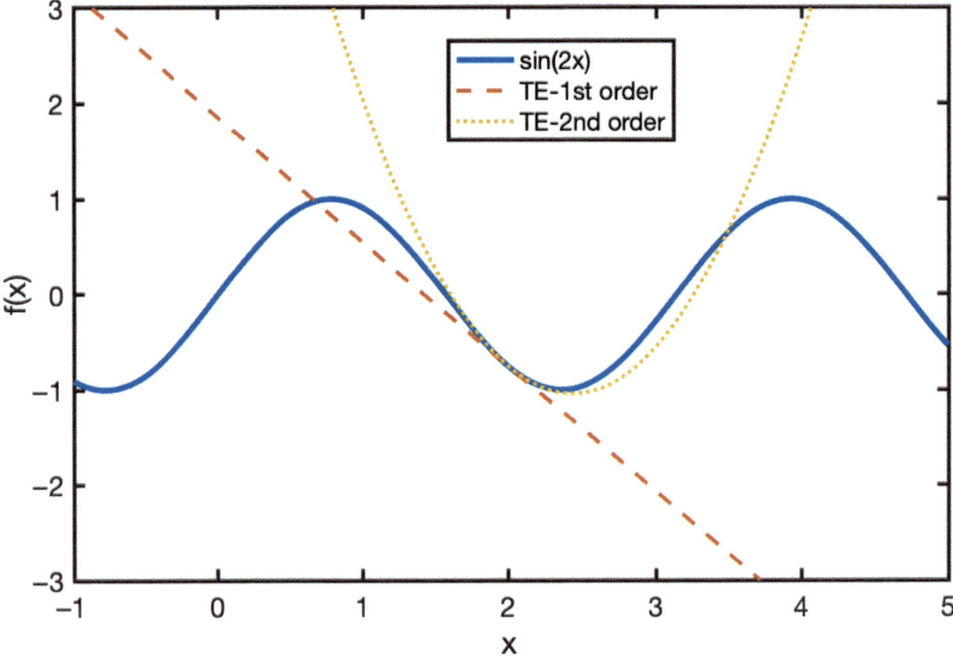

Figure 0.6 Adding now the second-order Taylor approximation of $\sin(2x)$ around $x = 2$. It corrects for the curvature of the function around $x = 2$.

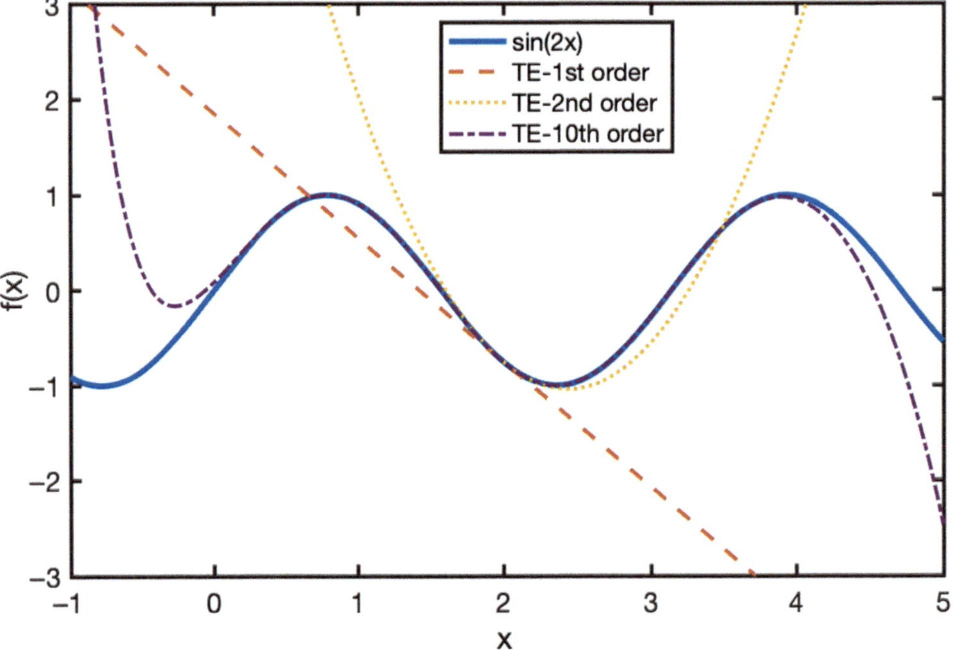

Figure 0.7 The tenth-order approximation provides a much better fit to the function between $0 \leq x \leq 4$.

This approach can be pursued indefinitely to get an exact value for $f(x)$ and it is called a Taylor infinite series:

$$\begin{aligned} f(x) &= f(x_0) + \frac{df}{dx}(x_0)(x - x_0) + \frac{d^2f}{dx^2}(x_0)\frac{(x - x_0)^2}{2} + \ldots + \frac{d^nf}{dx^n}(x_0)\frac{(x - x_0)^n}{n!} \\ &= \sum_{k=0}^{\infty} \frac{d^kf}{dx^k}(x_0)\frac{(x - x_0)^k}{k!}. \end{aligned} \quad (0.45)$$

We can see an important property of the Taylor series is that when estimating the value of a continuous function near a reference point x_0, if the distance $x - x_0$ is small, then the series can be truncated to include only a few low-order terms and remain somewhat accurate. This is an important point that we will use when discretizing differential equations into difference equations. For example, assuming that $x - x_0$ is small enough that all terms involving $(x - x_0)^k$ with $k \geq 2$ can be neglected, the truncated series gives us

$$f(x) = f(x_0) + \frac{df}{dx}(x_0)(x - x_0). \quad (0.46)$$

After reorganization, we get that when $x - x_0$ is small the average slope between x and x_0 approaches the value of the local slope at x_0

$$\frac{f(x) - f(x_0)}{x - x_0} \approx \frac{df}{dx}(x_0). \quad (0.47)$$

Again this would be true for a linear function (i.e., straight line) no matter what the distance between x and x_0, but as long as x and x_0 are sufficiently close to each other, it becomes a good approximation to any function (the qualifier sufficiently here is important!). However, the greater the magnitude of the higher-order derivatives of $f(x)$ at x_0, the smaller the gap between x and x_0 needs to be for this approximation to remain acceptable. This fundamental property of continuous functions, clearly laid out by the Taylor series, allows us to navigate between derivatives (right-hand side of Eq. 0.47) and differences (left-hand side of Eq. 0.47). For us, the Taylor series is the fundamental step to approximate derivatives with (finite) differences, that is, something like a portal that allows us to travel back and forth between these two worlds.

0.5.1 Exercises

- Compute the first three terms of the Taylor series for the following functions: (a) $\sin(x)$, (b) $\cos(2x)$, and (c) $x^2 + bx - c$ around the origin ($x_0 = 0$).
- Compute the first three terms of the Taylor expansion of the exponential function $f(x) = \exp(x)$ around the origin. Plot your results $f(x)$ and look at the quality of the approximation as your observation point x moves away from the reference point $x_0 = 0$. What could be the issue?
- Above we have used the Taylor series of $f(x)$ truncated to the first order to get an difference approximation for the derivative df/dx. Can you derive a similar approximation for the second-order derivative $d^2f/dx^2(x_0)$? (Hint: you may have to extend the truncation to the next order.)

Chapter 1

Introduction to multivariate calculus

1.1 Partial derivatives

So far we have focused on single variable functions $f(x)$ and introduced some basic concepts of calculus. While single variable calculus is relevant to ordinary differential equations (see Part II), many problems in Earth and planetary sciences depend on both time and space involving multivariate functions $f(x_1, x_2, \ldots)$ and require multivariate calculus. In this section, we first define the formalism related to differentiating multivariate functions and introduce some key concepts of multivariate calculus. Interested readers are encouraged to look at specialized textbooks to dig deeper in the subject.

We start with a function that depends on two variables $f(x_1, x_2)$ that we can identify, for example, with a topographic map (e.g., $f(x_1, x_2)$ is the elevation, x_1 and x_2 are two coordinates, pointing East and North). An example is shown in Figure 1.1a, and here is defined mathematically by

$$f(x_1, x_2) = 3\cos(x_1)x_2^2 - 2x_2\cos x_1 - 3. \tag{1.1}$$

Obviously this is not the most realistic landscape. The derivative of a single variable function is associated with its local slope, so how does that work when we consider more variables? In our example we can consider the slope of the topography along the coordinates x_1 or x_2, or any combination of the two. For simplicity, let us consider first the slope along each of the coordinates. If we consider traveling along the landscape according to a trajectory that is parallel to the coordinate x_1, that is, x_1 varies and x_2 is constant, then we can compute the local slope along this trajectory as

$$\frac{\partial f}{\partial x_1}(x_1, x_2 = \lambda) = \lim_{\delta x_1 \to 0} \frac{f(x_1 + \delta x_1, \lambda) - f(x_1, \lambda)}{\delta x_1}, \tag{1.2}$$

where λ is the constant value of x_2 along the trajectory. This mirrors the definition of the derivative for a single-variable function, but there are important points to note here. First, the symbol for this derivative is different; ∂x_1 refers to the partial derivative of the function $f(x_1, x_1)$ along the direction x_1, and implies that the other coordinate (x_2) is kept constant. It is different from $\frac{df}{dx_1}$, as we see later, where this derivative does not imply that x_2 is necessarily kept constant (e.g., it would vary here if x_1 depended on x_2). Using the definition of the partial derivative and our idealized topography, the slope along the x_1 direction at $x_2 = \lambda$ is given by

$$\frac{\partial f}{\partial x_1}(x_1, \lambda) = -3\lambda^2 \sin x_1 - 2\lambda \sin x_1. \tag{1.3}$$

Similarly we could travel across the landscape choosing a different path, specifically with $x_1 = \lambda$ a fixed value and varying x_2, and that would define the other partial derivative of $f(x_1, x_2)$

$$\frac{\partial f}{\partial x_2}(x_1 = \lambda, x_2) = \lim_{\delta x_2 \to 0} \frac{f(\lambda, x_2 + \delta x_2) - f(\lambda, x_2)}{\delta x_2}, \tag{1.4}$$

Introduction to Numerical Modeling in Earth and Planetary Sciences, Christian Huber, Oxford University Press.
© Christian Huber (2025). DOI: 10.1093/oso/9780198802716.003.0002

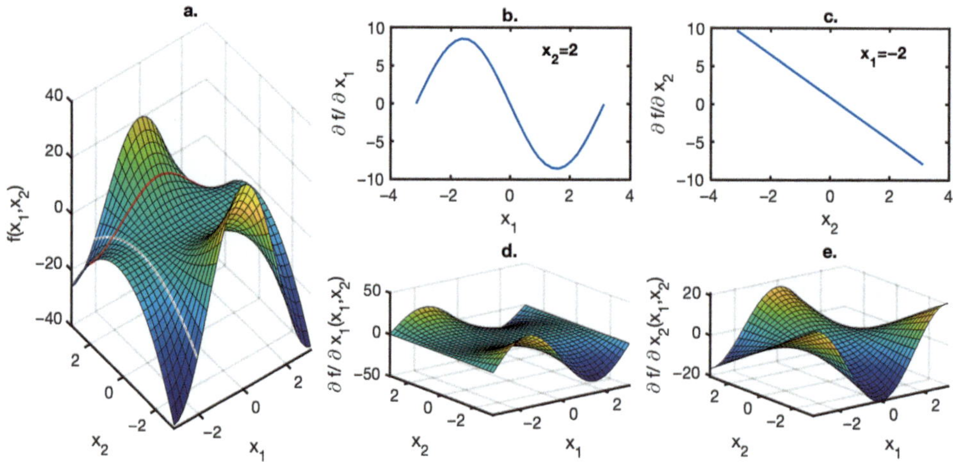

Figure 1.1 Example of a function f of two variables $f(x_1, x_2) = 3\cos(x_1)x_2^2 - 2x_2\cos x_1 - 3$ in Panel A. Two trajectories projected onto $f(x_1, x_2)$ are shown, one where x_1 is held constant and one where x_2 is held constant. Panels B and C show the partial derivative of the function along these trajectories. Panels D and E show the partial derivative functions over the whole domain.

and when applied to our function $f(x_1, x_2)$ defined above, it yields

$$\frac{\partial f}{\partial x_2}(\lambda, x_2) = 6\cos\lambda x_2 - 2\cos\lambda. \tag{1.5}$$

This function and the two partial derivatives at any position in the domain are shown in Figure 1.1. It shows that similarly to single-valued functions, where the derivative is itself a single-valued function, partial derivatives of a multi-variable function are also multi-variable functions (depending on the same variables). If these partial derivative functions are continuous they can be further differentiated to yield higher-order derivatives.

We defined the Taylor series expansion for single-variable functions in the previous chapter. Here we outline rapidly how this translates to multiple variables (for simplicity two variables x_1 and x_2) but the process generalizes to any dimensions. Let us first consider a reference point (x_1^0, x_2^0) where we can assume to know the value of the function and its partial derivatives (similar to what we did with single-variable functions). The change in the function between the reference position and an observation point (x_1, x_2) is

$$\Delta f = f(x_1, x_2) - f(x_1^0, x_2^0). \tag{1.6}$$

A possible and simple choice, given the definitions of the partial derivatives is to decompose the path between the two points into two segments that are both parallel to one of the variables, as shown in Figure 1.2. In that case, we can consider that, given one of the variables is fixed, the functions $f(x_1, \lambda)$ and $(f(\lambda, x_2)$ are two different single-variable functions and the formalism presented previously applies,

$$f(x_1, x_2) = \underbrace{f(x_1^0, x_2^0) + \frac{\partial f}{\partial x_1}(x_1^0, x_2^0)(x_1 - x_1^0)}_{\text{First segment}} + \underbrace{\frac{\partial f}{\partial x_2}(x_1, x_2^0)(x_2 - x_2^0)}_{\text{Second segment}} + \mathcal{O}(\Delta x_1 \Delta x_2, \Delta x_1^2, \Delta x_2^2), \tag{1.7}$$

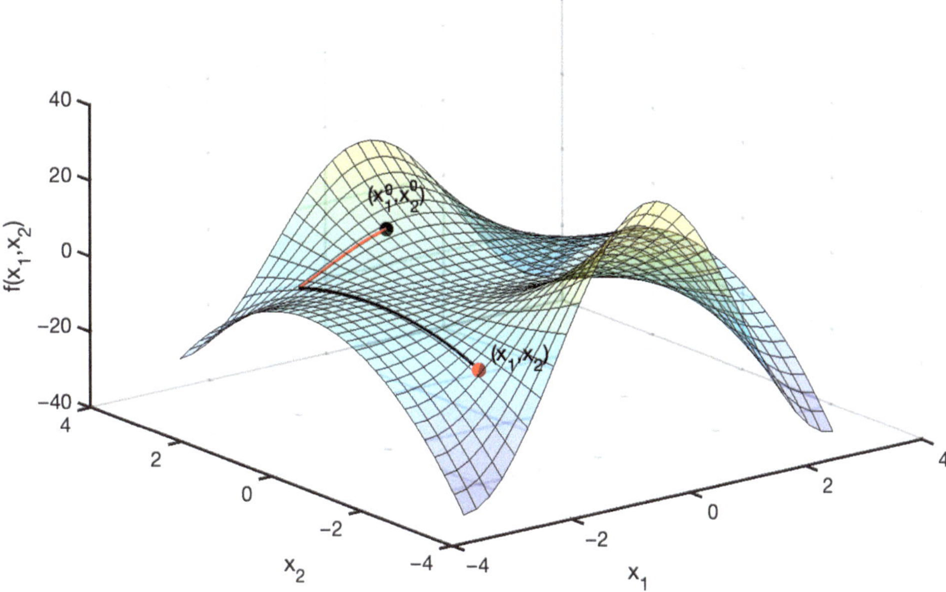

Figure 1.2 Illustration of the two (orthogonal) paths to go from position (x_1^0, x_2^0) to (x_1, x_2).

where, for example, $\mathcal{O}(\Delta x_2^2)$ means that terms that depend on the square of Δx_2 or higher powers are clumped together in that generic object (residual is of order 2 in this case). For conciseness, we define $\Delta x_1 = x_1 - x_1^0$ and $\Delta x_2 = x_2 - x_2^0$. The issue with the Taylor series above is that the slope (first-order partial derivative) with respect to x_2 is not taken at the reference point where information is assumed to be known, but at (x_1, x_2^0). An extra step is therefore required, if we consider $g(x_1, x_2) = \partial f / \partial x_2$ to be a continuous function; then we can, to the first order, apply a Taylor series expansion to it from the reference point (x_1^0, x_2^0) to correct for that

$$g(x_1, x_2^0) = g(x_1^0, x_2^0) + \frac{\partial g}{\partial x_1}(x_1^0, x_2^0)\Delta x_1 + \mathcal{O}(\Delta x_1^2). \tag{1.8}$$

Rewriting this equation in terms of $f(x_1, x_2)$ we get

$$\frac{\partial f}{\partial x_2}(x_1, x_2^0) = \frac{\partial f}{\partial x_2}(x_1^0, x_2^0) + \Delta x_1 \frac{\partial}{\partial x_1}\left(\frac{\partial f}{\partial x_2}(x_1^0, x_2^0)\right). \tag{1.9}$$

Inserting this back into Eq. 1.7

$$\begin{aligned} f(x_1, x_2) = f(x_1^0. x_2^0) &+ \frac{\partial f}{\partial x_1}(x_1^0, x_2^0)(x_1 - x_1^0) + \Delta x_2 \left(\frac{\partial f}{\partial x_2}(x_1^0, x_2^0) + \Delta x_1 \frac{\partial}{\partial x_1}\left(\frac{\partial f}{\partial x_2}(x_1^0, x_2^0)\right)\right) \\ &+ \mathcal{O}(\Delta x_1^2, \Delta x_2^2, \Delta x_1 \Delta x_2), \end{aligned} \tag{1.10}$$

which after reorganizing and recognizing that the last term is also second order with respect to Δx in general, we get

$$f(x_1, x_2) = f(x_1^0. x_2^0) + \frac{\partial f}{\partial x_1}(x_1^0, x_2^0)(x_1 - x_1^0) + \frac{\partial f}{\partial x_2}(x_1^0, x_2^0)(x_2 - x_2^0) + \mathcal{O}(\dots). \tag{1.11}$$

This is a natural extension of the first-order term of the Taylor series of single variable functions. Beware though that the number of terms involved in the Taylor series increases (nonlinearly) with the dimensionality for higher-order terms (involves cross-derivatives, like the one that we moved to $\mathcal{O}$ in Eq. 1.11).

1.1.1 **Exercises**

- Compute the first two orders of the Taylor series for the following functions: (a) $y^3 \sin(x)$; (b) $\cos(2x)\sin(xy)$; and (c) $yx^2 + bxy^3 - cx$ around the origin ($x_0 = 0$).

1.2 **Gradient**

An important object in calculus is the gradient of a continuous field $\phi(x, y)$ (for the simplicity of the argument, here we use a function that depends only on two variables) that involves partial derivatives of ϕ. If we want to compute the direction of the steepest slope of a function as well as its magnitude, for an arbitrary function, it is highly probable that the direction of the steepest slope is not parallel to the directions defined by the variables x_1 or x_2. The gradient provides the direction and magnitude of the steepest slope of a function at any given position, and it is a vector that is defined at any position to within the domain of the function, that is, it is a vectorial field $\vec{a}(x_1, x_2)$ that assigns a vector at each position (x_1, x_2). A vector can be decomposed into a sum of vectors, and one logical choice is to decompose it onto the basis of vectors that define the reference frame (e.g., unit vectors along the x_1 and x_2 axes, respectively denoted $\hat{x_1}$ and $\hat{x_2}$)

$$\vec{a} = a_1\hat{x_1} + a_2\hat{x_2}. \tag{1.12}$$

The coefficients (not vectors, but scalars) a_1 and a_2 are the projection of the vector $\vec{a}$ onto $\hat{x_1}$ and $\hat{x_2}$ (i.e., how much the vector extends in each of these directions; see Figure 1.3). We can consider a special vector field $\vec{a}$ where these scalar coefficients (projections onto $\hat{x_1}$ and $\hat{x_2}$) are the values of the partial derivative of the scalar function $\phi(x_1, x_2)$, namely, $a_1 = \frac{\partial\phi}{\partial x_1}$ and $a_2 = \frac{\partial\phi}{\partial x_2}$. This vector field $\vec{a}$ is then defined by

$$\vec{a} = \frac{\partial\phi}{\partial x_1}\hat{x_1} + \frac{\partial\phi}{\partial x_2}\hat{x_2}, \tag{1.13}$$

and this specific vector field is called the gradient of ϕ, it is generally denoted by $\nabla\phi$ and obviously depends on the scalar function ϕ. In more mathematical terms, the gradient is an operator that, in the particular case discussed here, takes a scalar field and yields a vector field, but we note that we can take the gradient of a vector field and obtain a higher order field (second-rank object like a matrix). In three dimensions a scalar field $\phi(x_1, x_2, x_3)$ admits a gradient

$$\nabla\phi = \frac{\partial\phi}{\partial x_1}\hat{x_1} + \frac{\partial\phi}{\partial x_2}\hat{x_2} + \frac{\partial\phi}{\partial x_3}\hat{x_3}. \tag{1.14}$$

As we discuss in Section 1.4, the orientation of the vectors of the gradient of ϕ are locally orthogonal to the trajectories where $\phi(x_1, x_2) = const$, or to use again the analogy with the topography of a landscape, the direction of the steepest slope locally (direction of the gradient) is perpendicular to contours of constant elevation. A last note before moving to other differential operators of fields is that the choice of reference frame (e.g., choice of directions for x_1 and x_2,

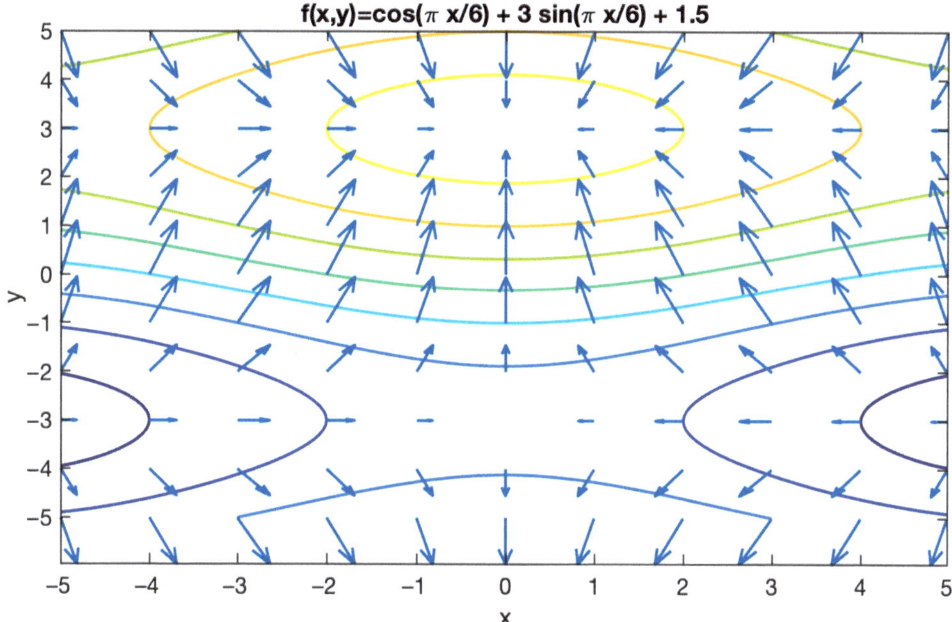

Figure 1.3 Lines of constant values of the function $f(x, y) = \cos(\pi x/6) + 3\sin(\pi x/6) + 1.5$ and its gradient field (arrows). Note that the gradient is locally always perpendicular to the lines of constant values of the function.

or the nature of the reference frame like Cartesian versus polar coordinates) informs how the gradient is expressed in this particular reference frame; however, the function alone dictates its overall direction and magnitude, and these are similar to different view points of the same object!

1.2.1 Exercises

- Compute the gradient of the function $f(x, y) = \cos(x) + \sin(y)$. For what position(s) (x, y) is the gradient component in the x-direction the greatest in magnitude ?
- In the previous example plot the $f(x, y)$ and the gradient at $y = 0$ and $y = 1$. How is the gradient oriented in each case ?

1.3 Divergence and curl

For simplicity we start with a problem involving a vectorial field that points everywhere in the same direction but where the magnitude depends only on one of the coordinates, here x (see Figure 1.4 for an example). Imagine that the vectorial field of interest $\vec{a}$ represents the velocity of an object (a car) moving along a rectilinear trajectory aligned with the coordinate x, but that the velocity along the path V_x can change along the trajectory. Imagine now that the car is modern and has front- and rearview cameras and you are the passenger. In the reference frame of the car, the pavement is coming towards you on the frontview camera and away from you on the rearview camera. The magnitude of the two velocities, inferred by the rate of movement of the pavement towards (front) and away (rear) from you is the same. If it was not, then the velocity of the front and rear of the car you sit in would be different. If the magnitude of the front velocity was greater than the rear velocity, then the front of the car would move forward with respect to the back of the car and the car's length would increase accordingly; conversely, if the magnitude of the front velocity is smaller than the back velocity, then you will find the car length would decrease accordingly. Obviously we expect the car not to change length (no relativist effect!) and the two velocities have to match. In this case we infer that the change in one-dimensional volume (length) l of the car is tied to the variation of its velocity field

$$\Delta l \propto \Delta V_x \approx \frac{\partial V_x}{\partial x} l, \tag{1.15}$$

where l is the original length of the object (car) at the edges of which the measurements are made and Δl is the change in length of the object. Dividing both sides by the length l to get a relative change in length and letting the length l be infinitesimally small and defining its rate of change with time $\dot{l}$, we get

$$\dot{\epsilon}_{xx} = \lim_{l \to 0} \frac{\Delta \dot{l}}{l} = \frac{\partial V_x}{\partial x}, \tag{1.16}$$

where we can recognize that the (longitudinal) strain-rate $\dot{\epsilon}_{xx}$ is equal (in this one-dimensional world) to the partial derivative (along x) of the velocity field V_x. So there is a connection between partial derivatives and the expansion or contraction expected from a vectorial field. We can show in three dimensions, assuming a general velocity field $\vec{V}(x_1, x_2, x_3)$ that is continuous, that the relative change in volume of the object $\Delta\mathcal{V}/\mathcal{V}$ is given by the sum of the contributions of the variations of the velocity along the different coordinates, and this generalizes to

$$\frac{\Delta \mathcal{V}}{\mathcal{V}} = \dot{\epsilon}_{x_1x_1} + \dot{\epsilon}_{x_2x_2} + \dot{\epsilon}_{x_3x_3} = \frac{\partial V_1}{\partial x_1} + \frac{\partial V_2}{\partial x_2} + \frac{\partial V_3}{\partial x_3}, \tag{1.17}$$

where V_j is the velocity component along the x_j direction. The sum of the partial derivatives in the last equality is defined as the divergence operator, here applied to the vectorial field $\vec{V}$, and it is denoted

$$\nabla \cdot \vec{V} = \frac{\partial V_1}{\partial x_1} + \frac{\partial V_2}{\partial x_2} + \frac{\partial V_3}{\partial x_3}. \tag{1.18}$$

Because the divergence of a vectorial field relates to whether the field induces contraction or expansion, the divergence of vectorial fields in physics is often tied to conservation statements (mass, momentum, or energy conservation). Note also that the divergence operator as we have seen here generates a scalar field (volume has no directionality in our example above) from a vectorial field. This is on stark (opposite) contrast to the gradient operator. The divergence operator cannot be applied to a scalar field, but it can be applied to higher-order fields, for example, the divergence of a stress field $\boldsymbol{\sigma}$ (tensor with nine components in three dimensions at each position) generates a vectorial force field

$$\nabla \cdot \boldsymbol{\sigma} = \vec{f}, \tag{1.19}$$

where $\vec{f}$ is the force field.

The divergence of three different two-dimensional vector fields is shown in Figure 1.4. The field $\vec{a}_1 = (1, 1)$ yields a divergence that is obviously zero, given both component of the field are homogeneous (do not depend on position). The second field $\vec{a}_2 = (c - \sqrt{x_1^2 + x_2^2}, c - \sqrt{x_1^2 + x_2^2})$ is shown in Figure 1.4b. This field is not homogeneous but is unidirectional. The bottom of Figure 4.1b shows the divergence of $\vec{a}_2$, a scalar field that varies over space depending on the distance of the point from the position (c, c). In this particular case, the finite divergence is caused by the varying magnitude (the direction of the field $\vec{a}_2$ is the same at any point) of the field with position. Figure 1.4c shows the vector field $\vec{a}_3 = (x_1^2, x_2^3)$ and its divergence; here both magnitude and direction of the vector field contribute to the divergence.

Another differential operator can be applied to vectorial fields (in three dimensions), the curl which describes the circulation in the field. It is generally written as

$$\nabla \times \vec{a} = \vec{b} \tag{1.20}$$

and it is a vectorial field that points in the direction orthogonal to the maximum local circulation (rotation) (right-hand rule for the sign and orientation of the vector) and its magnitude quantifies the amount of local circulation (circulation density). There are important examples of vectorial fields that are obtained through the curl of vectorial potential fields, for example, the vorticity in a fluid (curl of the fluid's velocity field) or the magnetic field (curl of a vector potential field). A vector field that is curl-free is called irrotational ($\nabla \times \vec{a} = 0$). Mathematically, the curl of vector $\vec{a}$

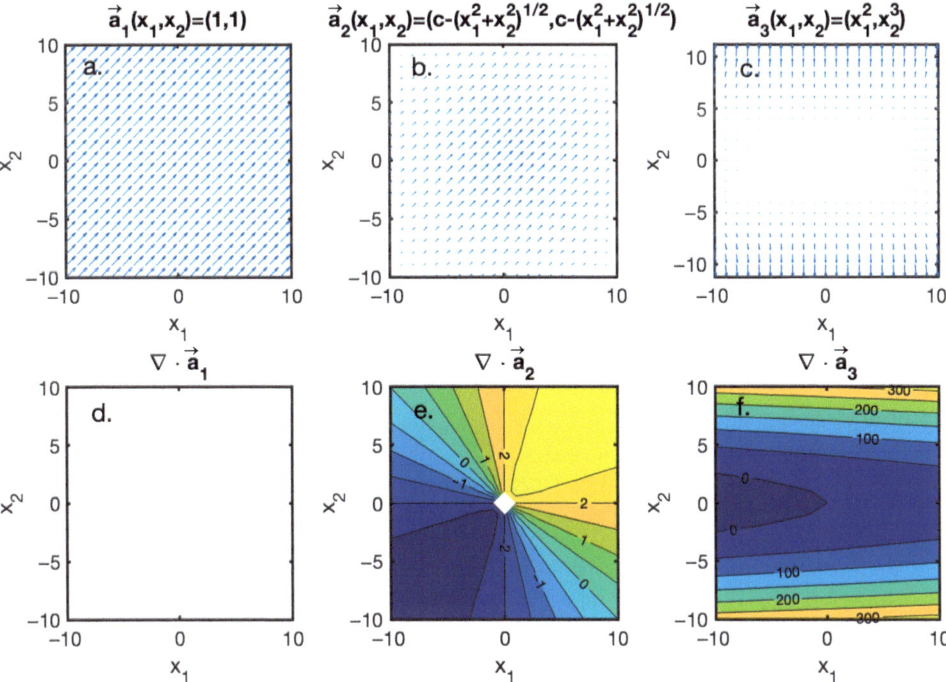

Figure 1.4 Three two-dimensional vector fields (top row) and their associated divergence scalar field (bottom row). The field $\vec{a}_1$ is solenoidal (divergence-free).

with components a_1, a_2, a_3 is defined, by component (here component i),

$$\left(\nabla \times \vec{a}\right)_i = \sum_{j=1}^{3}\sum_{k=1}^{3} \epsilon_{i,j,k} \frac{\partial a_k}{\partial x_j}, \tag{1.21}$$

where $\epsilon_{i,j,k}$ is the Levi–Civita symbol, which is defined as

$$\epsilon_{i,j,k} = \begin{cases} 1 & \text{if i, j, k is even permutation, e.g. (i, j, k)} = (1,2,3) \text{ or } (2,3,1) \text{ or } (3,1,2) \\ -1 & \text{if i, j, k is odd permutation, e.g. (i, j, k)} = (1,3,2) \text{ or } (3,2,1) \text{ or } (2,1,3) \\ 0 & \text{if indices are repeated.} \end{cases} \tag{1.22}$$

In Cartesian coordinates (x_1, x_2, x_3), the curl of the vector $\vec{a}$ is

$$\nabla \times \vec{a} = \begin{pmatrix} \frac{\partial a_3}{\partial x_2} - \frac{\partial a_2}{\partial x_3} \\ -\frac{\partial a_3}{\partial x_1} - \frac{\partial a_1}{\partial x_3} \\ \frac{\partial a_2}{\partial x_1} - \frac{\partial a_1}{\partial x_2} \end{pmatrix}. \tag{1.23}$$

An illustration of the curl of three vectorial fields is shown in Figure 1.5. The field $\vec{a}_1$ (Figures 1.5a, d) is unidirectional but not homogeneous and has a finite amount of rotation because of the variation in magnitude with position. The second field $\vec{a}_2$ is irrotational and actually describes a solid body rotation (no deformation of the field), which does not contribute a net circulation. Figures 1.5c, f ($\vec{a}_3$ and its associated curl) are also irrotational but actually have a net (positive) divergence (i.e., the divergence field of $\vec{a}_3$ is homogeneous).

Higher-order derivatives can be readily built from combinations of these vectorial calculus operators, as long they are consistent with one another in terms of the object to which they apply. For example, we cannot take the divergence or the curl of a scalar field, therefore an operator such as $\nabla \times (\nabla \cdot \vec{a})$ is not properly defined. There are some important vectorial operators and properties for second-order derivatives that are worth spending a little time on. We consider four cases and go over the groundwork required to demonstrate these properties.

1. $\nabla \cdot \left(\nabla \times \vec{a}\right) = 0$. We use the notation provided earlier to demonstrate that this operator applied to any continuous vectorial field $\vec{a}$ is null. First, reminding ourselves of the definition of the divergence operator we can write

$$\nabla \cdot \left(\nabla \times \vec{a}\right) = \sum_i \frac{\partial}{\partial x_i} \left(\nabla \times \vec{a}\right)_i . \tag{1.24}$$

We can expand the curl operator in the parenthesis in terms of the formalism introduced with the Levi–Civita symbol to get know

$$\nabla \cdot \left(\nabla \times \vec{a}\right) = \sum_{i,j,k} \frac{\partial}{\partial x_i} \left(\epsilon_{i,j,k} \frac{\partial}{\partial x_j} a_k\right). \tag{1.25}$$

It is important here to remember that $\epsilon_{i,j,k}$ is independent of the position and therefore can be moved in and out of spatial derivatives. Secondly, note that we consider the vectorial field $\vec{a}$ to

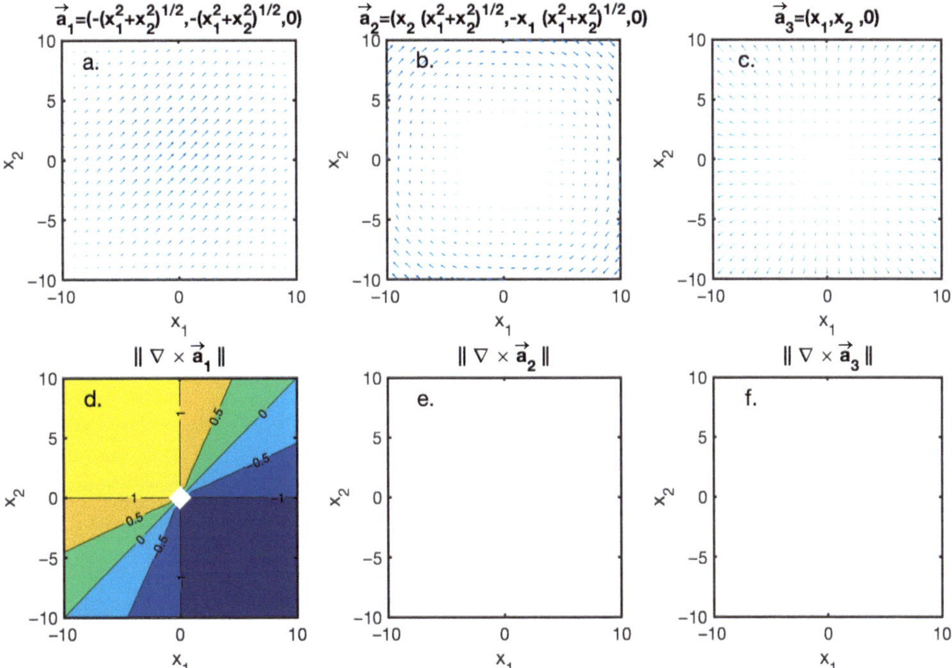

Figure 1.5 Three examples of "planar" three-dimensional vectorial fields (top row) and their associated curls (bottom). Fields $\vec{a}_2$ and $\vec{a}_3$ are irrotational (curl-free), specifically $\vec{a}_2$ is a solid body rotation around an axis centered at origin and extending in the third dimension x_3. Note: the vector field $\vec{a}_3$ has a constant divergence of 2 at any point (x_1, x_2, x_3).

be at least twice continuous, which means that switching the order of the derivative applied to it does not influence the result. In other words, differentiating it first with respect to x_j or x_i does not matter. So that allows us to write

$$\nabla \cdot (\nabla \times \vec{a}) = \sum_{i,j,k} \frac{\partial}{\partial x_j} \left(\epsilon_{i,j,k} \frac{\partial}{\partial x_i} a_k \right). \tag{1.26}$$

Using now that $\epsilon_{i,j,k} = -\epsilon_{j,i,k}$, we can write

$$\nabla \cdot (\nabla \times \vec{a}) = -\sum_{i,j,k} \frac{\partial}{\partial x_j} \left(\epsilon_{j,i,k} \frac{\partial}{\partial x_i} a_k \right), \tag{1.27}$$

which is the expanded version of

$$\nabla \cdot (\nabla \times \vec{a}) = -\sum_{i,j,k} \frac{\partial}{\partial x_j} \left(\epsilon_{j,i,k} \frac{\partial}{\partial x_i} a_k \right) = -\nabla \cdot (\nabla \times \vec{a}) \tag{1.28}$$

and the only one that is possible is for this operator to be null for any vectorial field $\vec{a}$.

2. $\nabla \times (\nabla f) = 0$. This states that the curl applied to any scalar field gradient (itself a vectorial field) is null. The demonstration here follows tightly to the previous case, and we use the antisymmetry of the Levi–Civita symbol once again. The first step is to write the curl in terms of the Levi–Civita symbol

$$\nabla \times (\nabla f) = \sum_{j,k} \epsilon_{i,j,k} \frac{\partial}{\partial x_j} \frac{\partial f}{\partial x_k}. \tag{1.29}$$

Again, we assume here that the scalar field f is at least twice continuous (i.e., can be differentiated twice and the order of the derivatives does not matter), which allows us to write

$$\nabla \times (\nabla f) = \sum_{j,k} \epsilon_{i,j,k} \frac{\partial}{\partial x_k} \frac{\partial f}{\partial x_j}. \tag{1.30}$$

Using the antisymmetry of $\epsilon_{i,j,k} = -\epsilon_{i,k,j}$ we get

$$\nabla \times (\nabla f) = -\sum_{j,k} \epsilon_{i,k,j} \frac{\partial}{\partial x_k} \frac{\partial f}{\partial x_j} = -\nabla \times (\nabla f), \tag{1.31}$$

which proves that the curl of any gradient is null.

3. $\nabla \times \left(\nabla \times \vec{a}\right) = \nabla \left(\nabla \cdot \vec{a}\right) - \nabla^2 \vec{a}$. The demonstration of this property requires a little more work but follows the same overall idea. First using the Levi–Civita symbol, we can write

$$\nabla \times \left(\nabla \times \vec{a}\right)_i = \sum_{j,k} \epsilon_{i,j,k} \frac{\partial}{\partial x_j} \left(\sum_{k,l,m} \epsilon_{k,l,m} \frac{\partial a_m}{\partial x_l} \right). \tag{1.32}$$

Knowing that the Levi–Civita symbol is independent of position we have

$$\nabla \times \left(\nabla \times \vec{a}\right)_i = \sum_{j,k} \sum_{l,m} \epsilon_{i,j,k} \epsilon_{k,l,m} \frac{\partial}{\partial x_j} \left(\frac{\partial a_m}{\partial x_l} \right). \tag{1.33}$$

The product of two ϵ is either 1 or −1 if the three indices are different, depending on the permutations. We can decompose all outcomes into positive and negative permutations to get

$$\nabla \times \left(\nabla \times \vec{a}\right)_i = \underbrace{\sum_{j,k} \epsilon_{i,j,k} \epsilon_{k,i,j} \frac{\partial}{\partial x_j} \left(\frac{\partial a_j}{\partial x_i} \right)}_{l=i,m=j} - \underbrace{\sum_{j,k} \epsilon_{i,j,k} \epsilon_{k,j,i} \frac{\partial}{\partial x_j} \left(\frac{\partial a_i}{\partial x_j} \right)}_{l=j,m=i}. \tag{1.34}$$

Reorganizing the terms and using the definition of the Levi–Civita symbol, we get

$$\nabla \times \left(\nabla \times \vec{a}\right)_i = \sum_{j,k} \frac{\partial}{\partial x_i} \left(\frac{\partial a_j}{\partial x_j} \right) - \sum_{j} \frac{\partial}{\partial x_j} \left(\frac{\partial a_i}{\partial x_j} \right) = \nabla \left(\nabla \cdot \vec{a}\right)_i - \nabla^2 \vec{a}_i. \tag{1.35}$$

4. $\nabla \cdot \left(\nabla \vec{a}\right) = \nabla^2 \vec{a}$. This operator is quite common in continuum mechanics and it is called the Laplacian of $\vec{a}$. In index notation we get

$$\nabla \cdot \left(\nabla \vec{a}\right)_j = \sum_{i} \frac{\partial}{\partial x_i} \left(\frac{\partial a_j}{\partial x_i} \right) = \sum_{i} \frac{\partial^2 a_j}{\partial x_i^2} = \frac{\partial^2 a_j}{\partial x_1^2} + \frac{\partial^2 a_j}{\partial x_2^2} + \frac{\partial^2 a_j}{\partial x_3^2}. \tag{1.36}$$

1.4 Gauss and Stokes theorems

There are two fundamental theorems that prove to be very useful when dealing with conservation laws in physics and vectorial fields. The first one is sometimes called the divergence theorem, Gauss's theorem, or even Green's theorem, and relates to the previous discussion about the significance of the divergence operator, that is, how it measures the contraction or expansion of a vectorial field. For any vectorial field $\vec{a}$, Gauss's theorem's states that

$$\int_V \nabla \cdot \vec{a}\, \mathrm{dV} = \oint_A \vec{a} \cdot \hat{n}\, \mathrm{dA}, \tag{1.37}$$

where $\oint_A$ symbolizes that we take the integral over a closed surface that wraps around the volume V. The theorem means that the expansion or contraction of the field $\vec{a}$ in a volume V is related with how much of vector field $\vec{a}$ enters (positive = expansion) or exits (negative = contraction) through the surface S of the volume. It is very convenient in many ways to derive conservation laws (where $\vec{a}$ will relate to a conserved quantity) as it relates the conserved quantity to flux across the domain.

The second theorem is often called the curl theorem or Stokes's theorem,

$$\int_A \nabla \times \vec{a} \cdot \hat{n}\, \mathrm{dA} = \oint_{\mathcal{L}} \vec{A} \cdot \mathrm{d}\tilde{\mathrm{l}}, \tag{1.38}$$

where A is a surface enclosed by the closed path $\mathcal{L}$. Stokes's theorem states that we can measure the swirling (curl) of a vectorial field $\vec{a}$ over a region by just averaging the value of the field over its boundary (measuring its rotation). The symbol $\oint_{\mathcal{L}}$ means that the integral is computed over a closed path $\mathcal{L}$.

1.5 Helmholtz decomposition

The vectorial properties demonstrated above, specifically

$$\nabla \cdot (\nabla \times \vec{a}) = 0 \tag{1.39}$$

and

$$\nabla \times (\nabla f) = 0, \tag{1.40}$$

suggest that the vector field $\vec{F}$ that derives from the curl of another vector field is a divergence-free field

$$\vec{F} = \nabla \times \vec{a} \quad \rightarrow \quad \nabla \cdot \vec{F} = 0. \tag{1.41}$$

These divergence-free fields are called solenoidal fields. An example of a solenoid field in mechanics is the magnetic field, which is derived from a vector potential through a curl operator.

Importantly, although a solenoid field is divergence-free, it is not curl-free, as

$$\nabla \times \vec{F} = \nabla \times (\nabla \times \vec{a}) = \nabla (\nabla \cdot \vec{a}) - \nabla^2 \vec{a}. \tag{1.42}$$

Another option for a vector field $\vec{F}$ is to be curl-free $\nabla \times \vec{f} = 0$. These fields are also called irrotational. Using the other property we can easily see that such fields can be derived from the gradient of scalar fields

$$\vec{F} = -\nabla \phi, \tag{1.43}$$

where the sign is by convention and ϕ is called a scalar potential. For example, the gravity force field is irrotational because it derives from a scalar gravitational potential.

The main idea behind the Helmholtz theorem is that any vectorial field $\vec{f}$ can be constructed from a sum of a solenoidal and irrotational field

$$\vec{F} = \nabla \times \vec{a} - \nabla \phi. \tag{1.44}$$

This theorem is not demonstrated here (demonstrations can easily be found in vectorial calculus textbooks). Decomposing a vectorial field such as displacement into these two components is very useful to study body wave propagation, where the solenoidal field (no volume change) is related to the propagation of shear waves while the irrotational field is related to the propagation of compressional waves.

1.5.1 **Exercises**

- Show that the field $\vec{F} = (3y, 3x, 0)$ is irrotational.
- Show that for $\vec{F} = (xy, -y^2, 0)$, $\nabla \cdot \nabla \times \vec{F} = 0$.

Chapter 2

Elements of complex calculus

2.1 Introduction to complex variables

Complex numbers have been introduced gradually into mathematics (and physics) since the sixteenth century, when mathematicians exerted great efforts in generalizing the solution (roots) of polynomial functions. The set of real numbers can easily be shown to be insufficient when solving for the root of even simple quadratic functions. For example, the quadratic equation

$$x^2 + x + 1 = 0 \tag{2.1}$$

admits two roots

$$r_{1,2} = \frac{-1 \pm \sqrt{-3}}{2}, \tag{2.2}$$

but none of the roots are real because $\sqrt{-1}$ is not defined in the set of real numbers $\mathcal{R}$. The set of real numbers was therefore complemented by complex numbers that generalize the definition of the nth root of a number to extend to negative values even for even orders n, because the imaginary number i is defined as $i = \sqrt{-1}$. Up to this point, complex numbers can be seen as an abstract construction to remedy this issue, but as discussed later, they are not limited to an abstract concept.

As a starting point, consider the set of natural numbers $\mathcal{N} = 0, 1, 2, 3, \ldots$ and represent them in a line (Figure 2.1).

Using the addition as the only existing operation we can see that unless we want to travel along that line always in the same direction (right), we must define negative numbers that can be added to move to the left. The existence of a new type of numbers that allows us, when added to any number in $\mathcal{N}$, to move to the left expands our universe to a new set of numbers

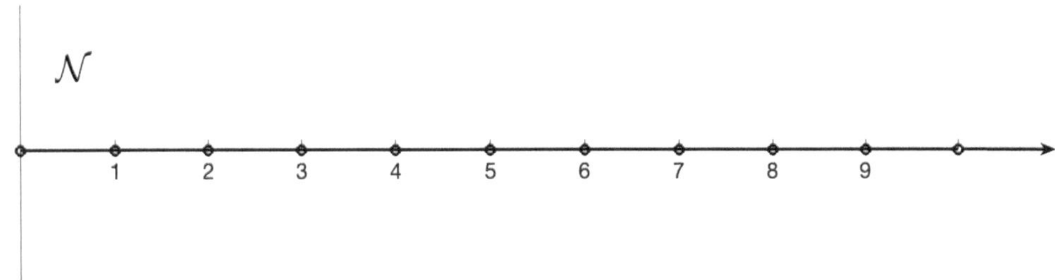

Figure 2.1 Illustration of the set of natural numbers $\mathcal{N}$; the line is open to the right but closed on the left at the origin.

Introduction to Numerical Modeling in Earth and Planetary Sciences, Christian Huber, Oxford University Press.
© Christian Huber (2025). DOI: 10.1093/oso/9780198802716.003.0003

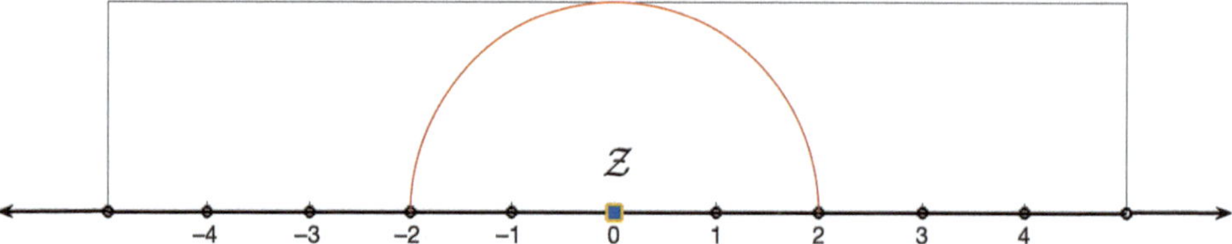

Figure 2.2 Illustration of the set $\mathcal{Z}$ of integer numbers (positive and negative) is open on both ends.

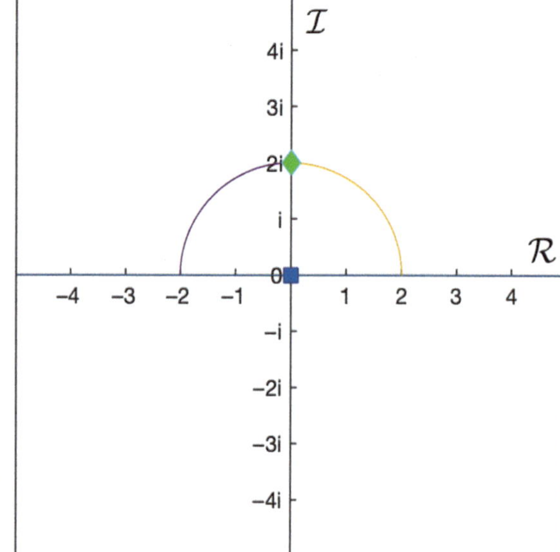

Figure 2.3 Illustration of the set of complex numbers. The real axis (horizontal) and imaginary axis (vertical) allow us to define uniquely any complex number as a combination of both.

$\mathcal{Z} = -\ldots, -2, -1, 0, 1, 2, \ldots$. As we consider the set of $\mathcal{Z}$ numbers as a line (infinite in both directions), we note some symmetry along the origin 0, where for a given value k along this set, $-k$ is equally distant from the origin but positioned in the opposite direction (see Figure 2.2). The symbol "-" therefore represents a symmetry operation with an "axis" defined at the position $x = 0$. A different (and perhaps at this time more confusing) way to see it is to regard the "-" symbol as a 180-degree rotation around the axis of rotation ($x = 0$). Two successive 180-degree rotations lands us in the original position $k = -(-k)$. All is well to this point, and we can understand the symbol "-" as a 180-degree (π in radian) rotation in our simplified framework (note that it works the same way for real numbers $\mathcal{R}$).

Now if we consider a similar approach and extend the realm of possibilities to other rotations (outside of 180 degrees), this will require a two-dimensional diagram. We can start with a 90-degree rotation around the origin, so half of the angle caused by the transformation $x \rightarrow -x$. Let us use a new symbol "i" for that 90-degree rotation (instead of "-" for 180 degrees) so that the transformation $x \rightarrow ix$ is a rotation of 90 degrees about the origin (the case for $2 \rightarrow 2i$ is shown in Figure 2.3). By construction, repeating twice the transformation to x such that $x \rightarrow ix \rightarrow i(ix)$ should be identical to a 180-degree rotation, and therefore

$$i(ix) = i^2 x = -x, \tag{2.3}$$

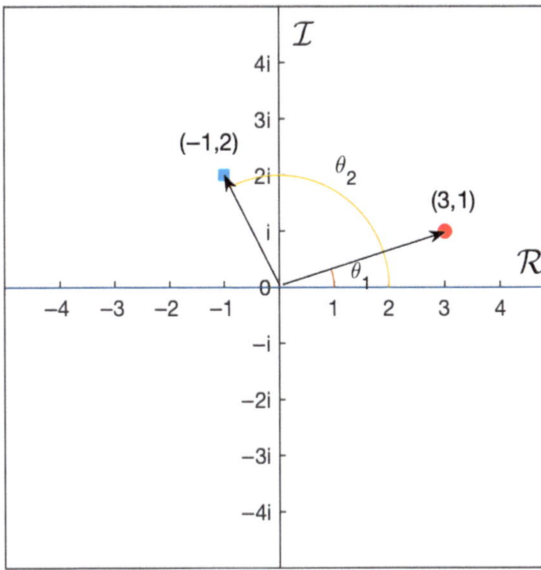

Figure 2.4 Multiplying a complex number by the imaginary number i amounts to a rotation of $\pi/2$ about the origin.

such that we have now a definition for the symbol i in terms of symbol "-" that states

$$i^2 = -1 \quad \text{or} \quad i = \sqrt{-1}. \tag{2.4}$$

The symbol i is generally referred to the imaginary number and the complex number formalism (extending the real set of numbers with the set of imaginary numbers ix) provides a means to define all the roots of polynomial functions such as the one of Equation 2.1

$$r_{1,2} = -\frac{1}{2} \pm \frac{\sqrt{3}}{2}i, \tag{2.5}$$

composed of a real part $(-1/2)$ and an imaginary part $(\pm\frac{\sqrt{3}}{2})$.

If multiplying a real number x by the imaginary number i is identical to a 90-degree rotation to the imaginary axis in Figure 2.3, we may wonder how to generalize the process to any rotation angle θ. Defining two complex numbers $c_1 = 3 + i$ and $c_2 = -1 + 2i$ (Figure 2.4), we can define them as two-dimensional vectors in our real+imaginary space and compute their respective angles with respect to the real axis. First, for generality let us assume that $a = 3, b = 1, c = -1, d = 2$ so that

$$c_1 = a + ib \tag{2.6}$$

$$c_2 = c + id. \tag{2.7}$$

Then the angle of these numbers (vectors) with respect to the real axis is

$$\theta_1 = \arctan\left(\frac{b}{a}\right) \tag{2.8}$$

$$\theta_2 = \arctan\left(\frac{d}{c}\right). \tag{2.9}$$

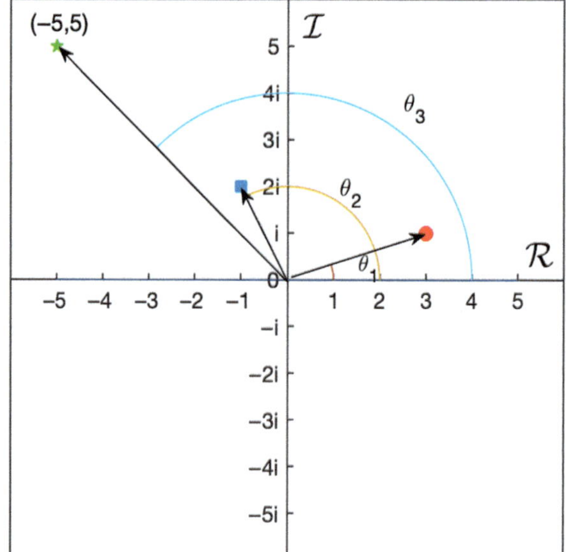

Figure 2.5 Taking the product of two complex numbers amounts to the sum of their respective angles with respect to the real axis and multiplying their magnitude.

Similarly, the Euclidean magnitude of these vectors is

$$\|c_1\| = \sqrt{a^2 + b^2} \tag{2.10}$$

$$\|c_2\| = \sqrt{c^2 + d^2}. \tag{2.11}$$

Let us now take the product $c_3 = c_1 c_2$ and compute the magnitude and orientation of that new vector in the real–imaginary two-dimensional world

$$c_3 = (a + ib)(c + id) = ac - bd + i(bc + ad), \tag{2.12}$$

so that the real part of c_3 $m = Real(c_3) = ac - bd$ and imaginary part is $n = Im(c_3) = bc + ad$ and

$$c_3 = m + in. \tag{2.13}$$

The magnitude of c_3 is then

$$\|c_3\| = \sqrt{(ac - bd)^2 + (bc + ad)^2} = \sqrt{(a^2 + b^2) * (c^2 + d^2)} = \|c_1\|\|c_2\|. \tag{2.14}$$

This result shows that the magnitude of the product of two complex numbers is the product of their magnitude, which is an important property we revisit shortly. The angle with respect to the real axis for c_3 is

$$\theta_3 = \arctan\left(\frac{bc + ad}{ac - bd}\right) = \theta_1 + \theta_2. \tag{2.15}$$

The last step requires some trigonometric operations, but the important point here is that the orientation of a complex number generated by the product of two complex numbers is exactly the sum of their respective orientation with further respect to a reference (here the real axis). Complex numbers therefore satisfy two important properties when multiplied:

$$\theta_3 = \theta_1 + \theta_2 \tag{2.16}$$

$$\|c_3\| = \|c_1\|\|c_2\|. \tag{2.17}$$

This clearly indicates that there is another convenient way to parameterize complex numbers outside of $c_1 = a + ib$, which directly portrays these properties. We discuss this in the next section.

2.2 Polar coordinates

The multiplicative properties of complex numbers indicate that a convenient way to represent them is by reference to polar coordinates

$$c_1 = \|c_1\| \exp(i\theta_1) = R_1 \exp(i\theta_1), \tag{2.18}$$

where R_1 is the magnitude of the complex number. This representation of complex numbers is called the polar coordinate system (R, θ) where a complex number is fully defined by its magnitude and angle with respect to the real axis. Although this notation is advantageous in many ways, it introduces a new concept: the exponential of a complex number $\exp(i\theta)$. We can use this polar coordinate system to make more intuitive sense of $\exp(i\theta)$ and consider the family of all complex number with magnitude $R = 1$. By definition, this family describes a unit circle in the complex plane (Figure 2.6) and each complex number on this circle is defined by its angle with respect to the real axis such that

$$c = \exp(i\theta). \tag{2.19}$$

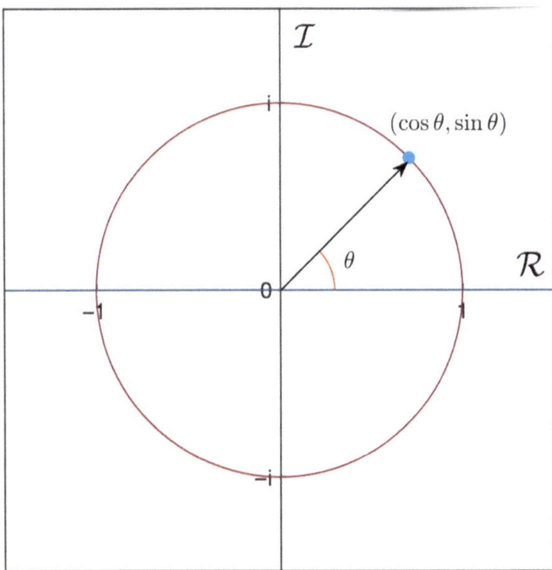

Figure 2.6 Illustration of the unit circle in complex space and how it relates to the Euler–Moivre equation $\exp(i\theta) = \cos(\theta) + i\sin(\theta)$.

From Figure 2.6 we can see that the position on the complex plane of each point is such that

$$Real(c) = \cos\theta \tag{2.20}$$

$$Im(c) = \sin\theta. \tag{2.21}$$

This amounts to the Euler–Moivre formula

$$\exp(i\theta) = \cos\theta + i\sin\theta. \tag{2.22}$$

2.2.1 **Exercises**

- What is the angle θ with respect to the real axis of the complex number $z_1 = -1 + i$? What is its magnitude ?
- If we want to rotate any complex number z_2 by $\pi/6$ around the origin without changing the magnitude of the vector, what complex number should we use to multiply z_2?
- Use polar coordinates to compute the four roots of the complex equation $z^4 = -1$.

Chapter 3

Elements of linear algebra

A common task in constructing numerical approximations to the solution of differential equations that describes physical processes of interest is the discretization of derivatives, which is given its proper emphasis in Part II of this book. Discretizing differential equations is the process by which differential equations are replaced by a set of algebraic equations that need to be solved together, and, in some cases, simultaneously. This transformation from ordinary differential equations (ODEs) or partial differential equations (PDEs) to algebraic equations is one of the fundamental steps of modeling (our third circle of Hell), and although it comes at the benefit of solving algebraic equations instead of differential ones, it is a main cause for the sometimes surprising results and headaches that come with numerical modeling.

For now, let us revisit some concepts of linear algebra that will become useful in our modeling adventures. A natural way to manipulate algebraic equations is through the use of objects such as matrices and vectors; this is where we will start. We then define or explain, with partial rigor, some of the common operations we will apply to matrices and vectors to make our modeling efforts more streamlined and efficient.

3.1 Vectors and matrices

A vector is a collection of items. It could be numbers, functions, or any collection of items one can combine by operations such as addition or product(s). The latter point is important mathematically, as vectors have certain properties that depend on how they transform, for instance, a collection of information (e.g., age, size, color of the eyes) for a group of people is not a true vector but just a list (array) of properties. There is no mathematically sensible transformation of this array that can be formed by the linear combination of these items:

$$\Omega = a_1(age) + a_2(size) + a_3(color\,of\,eyes) = ??? \tag{3.1}$$

Here we define vectors as arrays that are such that operations such as sum of its components are meaningful (linear transformations are possible and meaningful). For example, the three components of a velocity field are vectors, and the gradient of temperature or pressure is a vector.

We first review some simple properties of vectors before discussing their operations. First, the dimension of a vector is defined as the number of items that belong to it. For example, the velocity field in three dimensions can be decomposed into three independent components (v_1, v_2, v_3), and its dimension is 3. In this particular case each of the v_j $(j = 1, 2, 3)$ is a value (a scalar) that corresponds to the projected length of the vector along a reference axis. A two-dimensional example is provided for $\vec{v} = (v_x, v_y)$ in Figure 3.1. Here the reference axes are orthogonal (Cartesian x-y), where v_x is the length of the vector $\vec{v}$ projected onto the axis x and v_y is the equivalent for axis y. Two vectors with the same dimensions and defined over the same “space” (discussed in more details later) can be combined by addition or subtraction. In Figure 3.1b, two vectors (a_1, a_2)

Introduction to Numerical Modeling in Earth and Planetary Sciences, Christian Huber, Oxford University Press.
© Christian Huber (2025). DOI: 10.1093/oso/9780198802716.003.0004

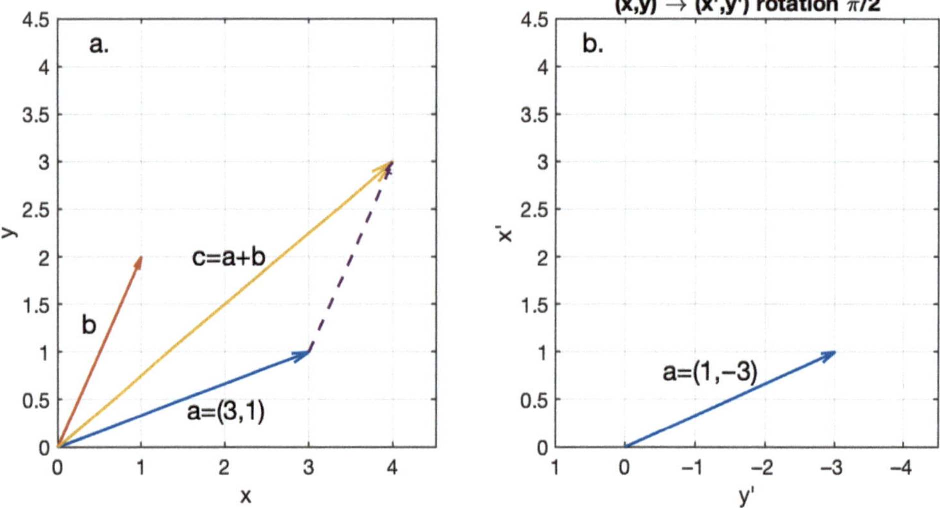

Figure 3.1 a): Illustration of the addition of two vectors $\vec{a}$ and $\vec{b}$. b): Representation of the same vector $\vec{a}$ but using a different choice of reference axes (x',y') rotated from (x,y) by $\pi/2$ counterclockwise.

and (b_1, b_2) are added to each other to form a new vector in that space (c_1, c_2). The rules for adding two vectors are simple: it is the addition of the component of each vectors $\vec{c} = (a_1 + b_1, a_2 + b_2)$. More generally, we can define a linear combination of vectors by

$$\vec{w} = \sum_j \alpha_j \vec{v}_j, \tag{3.2}$$

where α_js are scalar coefficients (any real number, for example) and $\vec{v}_j$ are vectors of the same dimension and defined in the same "space." Intuitively, we can guess that $\vec{w}$ has the same dimension as the vectors $\vec{v}_j$s. Another important property of vectors is their magnitude (discussed in the previous section about complex numbers). In a Euclidean space and Cartesian coordinates, the magnitude of a vector (its length) can be computed from Pythagoras theorem and

$$\|a\| = \sqrt{a_1^2 + a_2^2}, \tag{3.3}$$

which is the square root of the sum of the square of its components. Similarly $\|c\| = \sqrt{(a_1 + b_1)^2 + (a_2 + b_2)^2}$.

The choice of reference axes is somehow arbitrary, so what happens if we keep the same vector $\vec{a}$ but define different reference axes? In one case, the Cartesian axes (x-y) are used, and in the second case, the same object is written respective to the new set of axes (x'-y'), which are obtained by a rotation of $\pi/2$ counterclockwise from (x-y). Using the example illustrated in Figure. 3.1 we see that the same vector is expressed differently:

$$\text{In reference frame x} - \text{y} : \vec{a} = (3, 1) \tag{3.4}$$

$$\text{In reference frame x}' \text{-} \text{y}' : \vec{a} = (1, \text{-}3). \tag{3.5}$$

When expressing the coordinates of a vector, the set of reference axes needs to be defined, as the choice influences the coordinate's values. We call a set of reference axes that allow us to express any vectors in that space a vector basis. There is an infinite choice of possible vector basis (including any rotation or translation of the reference frame). Depending on the problem at hand, some choices of basis can be more adequate than others. One way to write a vector is as a linear combination over the vector basis (here normalized to unit vectors) with the linear coefficients being the coordinates of the vector in that basis. For example:

$$\vec{a} = a_1\hat{x} + a_2\hat{y} \tag{3.6}$$

$$\vec{a} = a_1'\hat{x'} + a_2'\hat{y'} \tag{3.7}$$

with $a_1, = 3$, $b_1 = 1$, and $a_1' = 1$, $b_1' = -3$. Changing the reference frame from (x,y) to (x',y') is a typical example of linear transformation of vectors. Let us call the transformation $f : \vec{v}_1 \rightarrow \vec{v}_2$ such that $f(\vec{v}_1) = \vec{v}_2$. If we consider now that f is the transformation that takes the original reference basis $\hat{x} = (1, 0)$ and $\hat{y} = (0, 1)$ into its rotated counterpart $\hat{x'}$ and $\hat{y'}$, then we consider f to be a counter-clockwise rotation of $\pi/2$ with $f(1, 0) = (0, 1)$ and $f(0, 1) = (-1, 0)$. If f is a linear transformation from one vector to another, it is possible to write it component by component, where $\hat{x}_i$ refers to component i of the vector $\hat{x}$

$$(f(\hat{x}))_1 = m_{11}\hat{x}_1 + m_{12}\hat{x}_2 = 0 \tag{3.8}$$

$$(f(\hat{x}))_2 = m_{21}\hat{x}_1 + m_{22}\hat{x}_2 = 1. \tag{3.9}$$

There are four coefficients m_{ij} in this transformation that need to be computed to fully define it. That means that an additional two equations are required for closure and we can use the effect of the same transformation on the other basis vector $\hat{y}$ to complement the set of equations

$$(f(\hat{y}))_1 = m_{11}\hat{y}_1 + m_{12}\hat{y}_2 = -1 \tag{3.10}$$

$$(f(\hat{y}))_2 = m_{21}\hat{y}_1 + m_{22}\hat{y}_2 = 0. \tag{3.11}$$

A convenient way to rewrite these expression is to define a matrix $\mathcal{M}_f$ such that

$$\mathcal{M}_f = \begin{pmatrix} m_{11} & m_{12} \\ m_{21} & m_{22} \end{pmatrix}, \tag{3.12}$$

such that

$$\hat{x'} = \mathcal{M}_f\hat{x} \tag{3.13}$$

$$\hat{y'} = \mathcal{M}_f\hat{y}. \tag{3.14}$$

where the four elements of the matrix can be solved from the equations above and $m_{11} = m_{22} = 0$ and $m_{12} = -m_{21} = -1$. We have now a full description of the transformation (counterclockwise rotation of $\pi/2$) through the matrix $\mathcal{M}_f$ and can transform any two-dimensional vector $\vec{a} = a_1\hat{x} + a_2\hat{y}$ into the rotated reference frame (basis x',y')

$$\mathcal{M}_f\vec{a} = a_1\mathcal{M}_f\hat{x} + a_2\mathcal{M}_f\hat{y} = a_1\hat{x}' + a_2\hat{y}' = -a_2\hat{x} + a_1\hat{y}. \tag{3.15}$$

Importantly, linear transformations applied to vectors can be written as matrix operators. This is true here and in general. The example shown here is simple and involves a matrix that takes as an input a vector $\vec{v}_1$ that has the same dimensions as the output vector $f(\vec{v}_1) = \vec{v}_2$, but the input and output vectors can be of different dimensions. We note that the number of rows and columns of the matrix $\mathcal{M}_f$ are respectively the dimensions of the number of elements of the output and input vectors. In the two-dimensional example of rotation shown above, it is possible to fully and uniquely describe the transformation matrix because a maximum of two independent vectors is needed to create any two-dimensional vector. In other words, although the choice of independent vectors to construct a vector basis of that two-dimensional world is infinite, there only can be two independent vectors that form that basis—no more and no less. Any third vector can be constructed from a linear combination (weighted sum) of the two basis vectors. Generally, in N dimensions, the number of linearly independent vectors that are needed to form a vector basis is N.

3.1.1 Exercises

- Build the 2×2 matrix for the transformation that sends the unit vector $\hat{x}$ into $(1\ 2)^T$ and $\hat{y}$ into $(-3\ 1)^T$.
- In three dimensions, build the matrix that rotates the vector by an angle of $\pi/6$ counterclockwise around $\hat{y}$.

3.2 Matrix inversion

In algebra classes, a common problem involves solving for the values of a vector $\vec{x}$ with a given matrix $\mathcal{M}$ and vector $\vec{R}$ such that

$$\mathcal{M}\vec{x} = \vec{R}. \tag{3.16}$$

As one quickly learns, there is a unique solution (set of elements for $\vec{x}$) only when the number of independent equations is identical to the number of unknowns. The number of unknowns here is the dimension of the vector $\vec{x}$ and the number of equations is given by the dimensions of the vector $\vec{R}$. If they do not match then either the solution is degenerate (not unique) or it does not exist. Only if the dimensions are identical can we expect a unique solution to exist. As we do with scalar equations such as

$$ax = r \quad \rightarrow x = a^{-1}r, \tag{3.17}$$

we are tempted to write with vector equations

$$\mathcal{M}\vec{x} = \vec{R} \quad \rightarrow \vec{x} = \mathcal{M}^{-1}\vec{R}. \tag{3.18}$$

This requires the inverse of a matrix $\mathcal{M}^{-1}$ to be defined and possible to calculate. Only square matrices can potentially be inverted (and not all can—see example in the exercises). One way to understand that is that if the matrix $\mathcal{M}$ represents a linear transformation of vectors, then $\mathcal{M}^{-1}$ is also a linear transformation, but one that satisfies

$$\mathcal{M}^{-1}\mathcal{M}\vec{v}_1 = \vec{v}_1 \tag{3.19}$$

where the combination of the two transformations is such that nothing happened to the vector (identity transformation). That means that any vector $\vec{v}_1$ transformed into a unique vector $\vec{v}_2$ can be transformed back into $\vec{v}_1$ without any ambiguity. So all the transformations must be uniquely defined (note that the two inverse transformations map each unique vector into another unique vector of the same dimension, which is generally called a bijection). As discussed, this is only possible if the dimensions of the input space and output space are the same and therefore that the matrices are square. Note that some of the square matrices cannot be inverted. For instance, take the linear transformation that projects any vector onto (for example) the x axis. Mathematically, one can define this projection transformation on any vector $\vec{a} = (a_x, a_y)$ as

$$p(\vec{a}) = a_x\hat{x} + 0\hat{y}. \tag{3.20}$$

Two vectors $\vec{a}_1 = (1, 1)$ and $\vec{a}_1 = (1, \lambda)$ for any value of λ will transform into the same output (1,0). The matrix for this transformation is

$$\mathcal{M}_p = \begin{pmatrix} 1 & 0 \\ 0 & 0 \end{pmatrix}, \tag{3.21}$$

and its inverse does not exist because of the problem of uniqueness of the transformation just discussed. Formally, a matrix $\mathcal{M}$ can be inverted if its determinant $det(\mathcal{M}) \neq 0$. The determinant $det(\mathcal{M}_p) = 1 \times 0 - 0 \times 0 = 0$.

3.2.1 Exercises

- What is the determinant of the two-dimensional rotation matrix around the origin (this matrix does not change the magnitude of a vector)?
- Building a transformation matrix in two dimensions that translates any vector by a fixed vector $\vec{\delta}$ is not trivial. If $\vec{a} = (a_x\ a_y)^T$ is a column vector in two dimensions, show that a translation transformation by $\vec{\delta} = (\delta_x\ \delta_y)^T$ amounts to perform the following linear transformations $\mathcal{P}\mathcal{M}\vec{b}$ where

$$\mathcal{P} = \begin{pmatrix} 1 & 0 & 0 \\ 0 & 1 & 0 \end{pmatrix}, \quad \mathcal{M} = \begin{pmatrix} 1 & 0 & \delta_x \\ 0 & 1 & \delta_y \\ 0 & 0 & 1 \end{pmatrix} \quad \vec{b} = \begin{pmatrix} a_x \\ a_y \\ 1 \end{pmatrix}. \tag{3.22}$$

3.3 Change of reference frame, matrix rotation

We discussed how there is an infinite choice of possible reference frames (or more appropriately here coordinate systems) to represent vectors. In this section, we first review how one expresses a vector in a different coordinate system (different vectorial basis) and then discuss how vectors (matrices) transform when different coordinate systems are used. Changing coordinate systems is quite common and sometimes simplifies significantly the algebra involved when dealing with a certain type of problems. It is important to remember that matrices are linear transformations that can apply to vectors, for example, a rotation, translation, or a combination of both.

Let us start with a vector $\vec{a}$ generically expressed in the standard coordinate system of unit vectors $\hat{e}_k$, such that in two dimensions, for example, $\hat{e}_1 = 1\hat{x} + 0\hat{y}$. The first important property of coordinate change is the transformation of a vector $\vec{a}$ from the standard coordinate system to a new set of basis vectors $\vec{v}_k$., and we write $[\vec{a}]_{v_k}$ as a way to identify the vector $\vec{a}$ written in this new basis. If we collect the basis vectors $\vec{v}_k$ as the columns of a matrix Ω, then

$$[\vec{a}]_{v_k} = \Omega^{-1}\vec{a}. \tag{3.23}$$

As an example, and not a proof, we define in two dimensions the vector $\vec{a} = 2\hat{e}_x - \hat{e}_y$. We use the new basis vectors

$$v_1 = \begin{pmatrix} 3 \\ 1 \end{pmatrix}, \quad v_2 = \begin{pmatrix} -1 \\ 3 \end{pmatrix}. \tag{3.24}$$

In this case, the coordinate change matrix is

$$\Omega = \begin{pmatrix} 3 & -1 \\ 1 & 3 \end{pmatrix}, \tag{3.25}$$

and its inverse is

$$\Omega^{-1} = \begin{pmatrix} 3 & 1 \\ -1 & 3 \end{pmatrix}. \tag{3.26}$$

Using the property written above, we get

$$[\vec{a}]_{v_k} = \Omega^{-1}\vec{a} = \begin{pmatrix} 1/2 \\ -1/2 \end{pmatrix}. \tag{3.27}$$

It is easy to show that

$$[\vec{a}]_{v_k} = \frac{1}{2}\vec{v}_1 - \frac{1}{2}\vec{v}_2, \tag{3.28}$$

as expected. In turning our attention to how matrices transform under changes in coordinate systems, we consider the following vector transformation

$$\vec{b} = \mathcal{A}\vec{x}, \tag{3.29}$$

where the matrix A is a generic transformation. Let us define

$$[\vec{b}]_{v_k} = \mathcal{B}[\vec{x}]_{v_k} \tag{3.30}$$

as the representation of this transformation in a new set of basis vectors $\vec{v}_k$. Using the property just described for the change in coordinates of a vector we get

$$\Omega^{-1}\vec{b} = \mathcal{B}\Omega^{-1}\vec{x}, \tag{3.31}$$

where the matrix of change in a coordinate system Ω has the new basis vectors as columns. Multiplying both sides to the left by Ω yields

$$\vec{b} = \underbrace{\Omega \mathcal{B} \Omega^{-1}}_{=\mathcal{A}} \vec{x} \tag{3.32}$$

and therefore

$$\mathcal{A} = \Omega \mathcal{B} \Omega^{-1}, \tag{3.33}$$

or equivalently

$$\mathcal{B} = \Omega^{-1} \mathcal{A} \Omega. \tag{3.34}$$

Considering now a simple example of matrix transformation upon coordinate change, we consider a simple rotation of $\pi/2$ in two dimensions and two independent matrices $\mathcal{A}$ and $\mathcal{B}$ expressed in standard coordinates as

$$\mathcal{A} = \begin{pmatrix} 2 & -1 \\ 1 & 2 \end{pmatrix}, \quad \mathcal{B} = \begin{pmatrix} 2 & 1 \\ 1 & 2 \end{pmatrix}. \tag{3.35}$$

The rotation matrix and its inverse (which is also its transpose) are

$$\mathcal{R} = \begin{pmatrix} 0 & -1 \\ 1 & 0 \end{pmatrix}, \quad \mathcal{R}^{-1} = \begin{pmatrix} 0 & 1 \\ 1- & 0 \end{pmatrix}. \tag{3.36}$$

The matrices $\mathcal{A}$ and $\mathcal{B}$ in the new coordinate system (rotated by $\pi/2$) are

$$\widetilde{\mathcal{A}} = \mathcal{R}^{-1} \mathcal{A} \mathcal{R} \tag{3.37}$$

$$\widetilde{\mathcal{B}} = \mathcal{R}^{-1} \mathcal{B} \mathcal{R}. \tag{3.38}$$

After some algebra we get

$$\widetilde{\mathcal{A}} = \begin{pmatrix} 2 & -1 \\ 1 & 2 \end{pmatrix}, \quad \widetilde{\mathcal{B}} = \begin{pmatrix} 2 & -1 \\ 1- & 2 \end{pmatrix}. \tag{3.39}$$

The change in coordinate system interestingly has not affected the numerical values in matrix $\mathcal{A}$ but has affected $\mathcal{B}$. This means that not all matrices behave identically upon coordinate change and the challenge is to identify a certain transformation (matrix) once we consider a different basis, yet it has to be the same overall transformation. This raises the question of how one identifies two matrices to encode the same transformation once written in different coordinate systems. This then leads us to the notion of matrix fingerprinting and (more specifically) to eigenvalues and eigenvectors (this is only valid for square matrices, i.e., matrices that transform a vector to another vector that exists in the same (vectorial) space).

3.3.1 Exercises

- Do the two following matrices commute, that is, $\mathcal{M}_1\mathcal{M}_2 \stackrel{?}{=} \mathcal{M}_2\mathcal{M}_1$ with

$$\mathcal{M}_1 = \begin{pmatrix} 1 & 1 \\ 0 & 1 \end{pmatrix}, \quad \mathcal{M}_2 = \begin{pmatrix} 1 & -1 \\ 1 & 1 \end{pmatrix}. \tag{3.40}$$

- Show that the product of the following matrices commutes

$$\mathcal{M}_3 = \begin{pmatrix} \lambda_1 & 0 \\ 0 & \lambda_2 \end{pmatrix}, \quad \mathcal{M}_4 = \begin{pmatrix} \lambda_3 & 0 \\ 0 & \lambda_4 \end{pmatrix}. \tag{3.41}$$

3.4 Eigenvalues and eigenvectors

A matrix operates a defined (linear) transformation on vectors, and regardless of the coordinate system the transformation should achieve the same results as it transforms a vector into another. However, we saw that a given change of coordinate system affects the matrices' representation in a different way. Some properties of the matrix remain unchanged under coordinate system change. For example, the trace of the matrices $\mathcal{A}$ and $\mathcal{B}$, that is, the sum of their diagonal elements, is not modified by the change in coordinate system (rotation). We show here how a coordinate change does not affect other properties of the transformation matrix. Note that every square transformation matrix admits a set of vectors $\vec{v}_j$ that are linearly independent from one another and that transform in a very simple way, that is, these special vectors are only rescaled or reversed (direction) upon transformation so that

$$\mathcal{A}\vec{v}_j = \lambda_j\vec{v}_j, \tag{3.42}$$

where λ_j is a scalar value that can be real or complex. λ_j is the scaling factor of the transformation for the vector $\vec{v}_k$. This conjugate pair of scaling factor and special vectors are called *eigenvalue-eigenvectors* of the matrix $\mathcal{A}$. An eigenvector is a special vector that transforms through the matrix $\mathcal{A}$ in a very simple and specific way, and its magnitude is scaled by the associated eigenvalue; it may change direction (sign) but not orientation (i.e., if the eigenvalue is real).

The number of linearly independent eigenvectors associated with the transformation matrix is equal to the dimension of the matrix if the latter can be inverted, and therefore it is possible to build a vector basis for the space upon which the transformation operates from eigenvectors alone. This is often convenient and we resort to it a few times in this book. Another important property of eigenvectors is that any two eigenvectors that have different eigenvalues are linearly independent. For example, assume (ad absurdum) that the eigenvector $\vec{v}_1$ can be expressed as a linear combination of the other eigenvectors $\vec{v}_{j\neq 1}$

$$\vec{v}_1 = \sum_{j\neq 1} a_j\vec{v}_j. \tag{3.43}$$

In this case, we can apply the transformation matrix $\mathcal{A}$ to $\vec{v}_1 - \sum_{j\neq 1} a_j\vec{v}_j$ and should by definition get the null vector. Yet,

$$\mathcal{A}\left(\vec{v}_1 - \sum_{j\neq 1} a_j\vec{v}_j\right) = \lambda_1\vec{v}_1 - \sum_{j\neq 1} a_j\lambda_j\vec{v}_j, \tag{3.44}$$

where the λs are the associated eigenvalues. Reorganizing the right side of the equation leads to

$$\vec{v}_1 - \sum_{j\neq 1} a_j \frac{\lambda_j}{\lambda_1} \vec{v}_j = 0, \tag{3.45}$$

which, by comparison with (Eq 3.43), shows that the only possible way to make that work would be if $\lambda_1 = \lambda_j$ for all values of j. So only eigenvectors sharing the same eigenvalues can possibly be linearly dependent.

The properties of the eigenvectors with respect to their transformation (matrix) suggests that they preserve some properties when the matrix is subjected to a change in coordinate system. We saw that vectors are modified by changes in a coordinate system and that will be true of eigenvectors, although a change in coordinate system should not affect the transformation that takes place when the matrix/transformation is applied to a given (specified) vector. For instance, if an eigenvector is rescaled by a factor λ through the transformation described by matrix $\mathcal{A}$, then a change in coordinate system should still find that the same vector (expressed in the new coordinate system) is rescaled by the same factor λ; otherwise the transformation before and after coordinate change are not similar. In other words, eigenvalues are independent of a change in coordinate system and eigenvectors just transform as one would expect (see discussion above). This is easy to demonstrate. Let us first assume that $\vec{w}$ is an eigenvector of the matrix $\mathcal{A}$ with associated eigenvalue λ

$$\mathcal{A}\vec{w} = \lambda\vec{w}. \tag{3.46}$$

Considering a change in coordinate system (vectors in square brackets describe the new coordinate system) so that

$$[\vec{w}] = \Omega^{-1}\vec{w}, \tag{3.47}$$

and in terms of matrix, as we have seen the transformation similar to $\mathcal{A}$ in the new coordinate system is

$$\mathcal{B} = \Omega^{-1}\mathcal{A}\Omega. \tag{3.48}$$

Then we can compute

$$\mathcal{B}[\vec{w}] = \Omega^{-1}\mathcal{A}\underbrace{\Omega\Omega^{-1}}_{identity}\vec{w} = \Omega^{-1}\mathcal{A}\vec{w} = \lambda\Omega^{-1}\vec{w} = \lambda[\vec{w}]. \tag{3.49}$$

This shows that eigenvalues remain unchanged with coordinate change and therefore provide another means to identify whether two matrices are similar (i.e., perform the same transformation). We can therefore use eigenvectors as a possible basis for the vectorial space on which the matrix operates and have the benefit that the eigenvalues are not affected by any change of coordinate system. This is a powerful property that we use when solving differential equations and when discussing stability of numerical schemes in future chapters.

So far we have not discussed how to compute eigenvectors and eigenvalues for a given matrix. The procedure is relatively simple and tractable for 2×2 matrices, but (in principle) the method shown here applies to any dimension. Starting from the matrix $\mathcal{A}$, we use Eq. 3.42 and reorganize it to get the vectorial equation

$$\mathcal{A}\vec{v} - \lambda\vec{v} = 0. \tag{3.50}$$

The second term remains unchanged when multiplied (to the left) by the identity matrix $\mathcal{I}$ (1 s on the diagonal and zeros everywhere else) which allows us to write

$$(\mathcal{A} - \lambda\mathcal{I})\vec{v} = 0. \tag{3.51}$$

The term in parenthesis is a matrix that applies to the eigenvector of $\mathcal{A}$ $\vec{v}$. An eigenvector cannot be null so there must exist two solutions to this linear problem: the trivial solution $\vec{v} = 0$ and the one we are interested in, where $\vec{v} \neq 0$. This means that the mapping of this new matrix $\mathcal{A}$-$\lambda\mathcal{I}$ is not unique and therefore that this matrix is not invertible, that is, its determinant must be 0. The determinant is

$$(a_{11} - \lambda)(a_{22} - \lambda) - a_{12}a_{21} = 0. \tag{3.52}$$

This quadratic equation (in λ) can be solved for (if we now have the matrix $\mathcal{A}$) to retrieve two eigenvalues (they may be complex). To find the two associated eigenvectors requires solving Eq. 3.42 twice, once for each eigenvalue. The choice of eigenvector is not unique (magnitude of the vector) so we can assume an eigenvector of the form $\vec{v}_1 = (v_{1x}\ 1)^T$ for example and then $\vec{v}_2 = (1\ v_{2y})^T$ for the second and solve for v_{1x} and v_{2y}.

3.4.1 **Exercises**

- Compute the eigenvalues and associated eigenvectors of the matrices, and you will see that the fact that they do not commute implies that their two eigenvectors are not similar.

$$\mathcal{M}_1 = \begin{pmatrix} 1 & 1 \\ 0 & 1 \end{pmatrix}, \quad \mathcal{M}_2 = \begin{pmatrix} 1 & -1 \\ 1 & 1 \end{pmatrix}. \tag{3.53}$$

Chapter 4

Treating functions as vectors

4.1 What is a vectorial space?

The notion of a vectorial space is central to linear algebra and to the rules behind vector sum and products. Given that this textbook does not aim at replacing mathematics textbooks and just presents a quick and informal overview of useful mathematical tools, we introduce here some simple notions without feeling the need for a more formal introduction. This section also serves an important point: the concept of what is and is not a vector extends well beyond the general idea of a vector being a set of numbers that has a direction and magnitude. We define vectorial spaces so as to expand our intuition beyond that simplistic view of what vectors are.

When defining a vectorial space, we need to define two sets, one possibly multi-dimensional, that is, the set $\mathcal{W}$ where the vectors exist, and one with a single dimension (numbers here called scalars) that we refer to as $\mathcal{S}$. When discussing an element of each set, for example, we refer to the element (vector) $V \in \mathcal{W}$ or the scalar $s \in \mathcal{S}$. For most applications, the set $\mathcal{S}$ will coincide with the set of real or complex numbers, whereas on the other end, the set $\mathcal{W}$ can be at first surprising, as we will discuss later. The space $\mathcal{W}$ can be of finite (three dimensions for our everyday physical space) or infinite dimensions.

A vectorial space must satisfy a list of mathematical properties:

1. there exists an operation to sum (add) vectors within that space. If $v_1, v_2 \in \mathcal{W}$, then there is a $v_3 \in \mathcal{W}$ such that $v_3 = v_1 + v_2$.
2. there is a neutral vector (element) in the space for the addition, where we use the notation 0, remembering that it may be more than a number (multidimensional) so that $v = 0 + v$.
3. the sum operation commutes, that is, the order of the addition does not matter, $v_1 + v_2 = v_2 + v_1$.
4. the sum operation is associative, that is, $(v_1 + v_2) + v_3 = v_1 + (v_2 + v_3)$.
5. the sum operation admits an inverse vector $-v_1$. So for every vector $v \in \mathcal{W}$ there is $-v \in \mathcal{W}$.
6. there exists a multiplication with a scalar operation such that $v_2 = s_1 v_1$ with $s \in \mathcal{S}$ and $v_2 \in \mathcal{W}$.
7. there is a neutral element in $\mathcal{S}$ called 1 such that $v = 1v$.
8. distributivity with multiplication, $s(v_1 + v_2) = sv_1 + sv_2$, and $(s_1 + s_2)v_1 = s_1 v_1 + s_2 v_1$.
9. $s_1(s_2 v_1) = (s_1 s_2)v_1$.

Again, this is not a very formal definition, but it will do for the moment. You may wonder how this will help us with the numerical solution of differential equations. The point is that linear algebra will be useful for us and the notion of vectorial spaces is tightly connected to it. Moreover, stability analysis of a numerical schemes often requires some knowledge that we will borrow from this section.

Introduction to Numerical Modeling in Earth and Planetary Sciences, Christian Huber, Oxford University Press.
© Christian Huber (2025). DOI: 10.1093/oso/9780198802716.003.0005

We see from the definitions above that the review of linear algebra in a previous section complies with the notion of vectorial space. What may come next is perhaps more surprising, but it will prove valuable in the context of numerical modeling.

4.2 Vectorial space of single variable continuous functions

We start with the claim that single variable continuous functions can be seen as vectors of an infinite dimension vectorial space. It may seem odd to treat functions as vectors, but given the definitions of what constitutes a vectorial space, we can first test whether this assumption is valid. If $f(x)$ and $g(x)$ are continuous (real-valued) functions, their sum $h(x) = f(x) + g(x)$ is continuous as well, that is, for any value of x_0:

$$\lim_{x\longrightarrow x_0} h(x) = \lim_{x\longrightarrow x_0} f(x) + \lim_{x\longrightarrow x_0} g(x) = f(x_0) + g(x_0) = h(x_0). \quad (4.1)$$

Thus $h(x)$ is also a continuous function and therefore lives in the same (vectorial) space. This is a good start. Now, for any $f(x)$ continuous function, we can find an opposite function $-f(x)$ such that $f(x) + (-f(x)) = 0$ is the null function (gives 0 for any value of input x), which is also the neutral vector for the addition. We can also easily show that the sum of two vectors commutes and that when considering three vectors the sum is associative. Multiplying a continuous function by a scalar (real constant, for example) does yield a continuous function and all the properties of vectorial fields attached to the product rules can be easily demonstrated. There is therefore no doubt, the space of continuous functions is a vectorial space! Perhaps a more complicated concept is the dimension of that space. We see later that the vectorial space of continuous functions has infinite dimensions. This can be problematic, but we can circumvent it by the fact that we are seeking for approximated, rather than exact solution, as we discuss later. The main question remains: why do we care about the vectorial space of continuous functions? The reason is rather simple: we seek to solve differential equations that describe the evolution of systems relevant to physical sciences, more specifically to Earth and planetary sciences. The mere assumption that differential equations can describe these phenomena implies that the functions that we aim to solve are smooth and continuous (at least within the domain of interest). This somehow restricts the family of functions that can be solutions to our differential equations to continuous functions, while some of these functions will involve multiple variables (see Part III devoted to PDEs). The overall discussion here applies to single-valued or multivariate functions equally well. For the sake of simplicity, here we review material that applies to the vectorial space of single-valued continuous functions.

4.3 Considering different basis functions

If continuous functions form a vectorial space, we can consider different choices of basis vectors (here, basis functions, which are continuous), that would allow us to span the whole space and represent any continuous function. Any continuous function can be decomposed into a set of basis functions in the same way as vectors in space can be decomposed into a basis of vectors, for example, Cartesian, cylindrical, or spherical coordinates. The main differences are that the vectors here are functions and that the number of elements in the basis is infinite. The first example goes back to a Taylor series and shows that it can be viewed as a representation of continuous functions as a series of polynomials.

4.3.1 Taylor series

The Taylor series was introduced in Section 0.5. Taylor series for a continuous function $f(x)$ is defined as

$$f(x) = \sum_{k=0}^{\infty} \frac{d^k f}{dx^k}(x_0)\frac{(x - x_0)^k}{k!}. \tag{4.2}$$

It is an infinite series ($k = 0, \ldots, \infty$), where the terms $\frac{d^k f}{dx^k}(x_0)$ are merely coefficients that we can write as $a_k k!$, which allows us to recast it as

$$f(x) = \sum_{k=0}^{\infty} a_k (x - x_0)^k. \tag{4.3}$$

The polynomials of order $k = 0, \ldots, \infty$ are each continuous functions and as discussed, a Taylor series can represent any continuous function, so these infinite sets of polynomials span the vectorial space and therefore are a suitable basis for it.

4.3.2 Exercises

Compute the first four terms of the Taylor series of the following function using $x_0 = 0$:

- $\cos(2x)$
- $\sin(2 * pix/L)$
- $\sin x/x$.

4.3.3 Cosine and sine series representation

We use now a more useful choice of basis to look into the nature of solutions of differential equations (dynamic or numerical stability) and consider periodic waves with different frequencies as a set of basis functions. The first two choices are restrictive in that they only work for a subset of continuous functions, but they are complementary and together they span the whole vectorial space of continuous functions. First, we introduce odd and even continuous functions and discuss them individually. Even functions are such that any one of them (here, $f(x)$) is such as that they are "symmetrical" around the origin

$$f(-x) = f(x). \tag{4.4}$$

The $\cos x$ function is even, as is the absolute value function $f(x) = x^2$. Alternatively, functions that are antisymmetrical around the origin

$$f(-x) = -f(x) \tag{4.5}$$

are called odd functions (for example, $f(x) = x, \sin x$). Importantly, any continuous function can be constructed from the sum of even and odd functions. Because of boundary conditions of the differential equation of concern, if we consider that the solution is by construction odd or even, we can restrict the infinite vectorial space of continuous functions to those that are odd or even, that is, a subset of the vectorial space, but still infinite in dimension. The advantage is that this restriction offers a fairly simple choice of vectorial basis. We start with even functions. As mentioned. $\cos x$ is an even function, $\cos(-x) = \cos x$. The idea here is to use a series of cosine functions as a basis for even continuous functions in a way similar to how we use polynomials in a Taylor series. In the latter it was the order of the power that distinguished the different vectors of the basis, whereas in the case of the cosine series representation, it is the frequency (over space or time) of the waves represented by the cosine functions that distinguishes the different vectors of the basis. Considering a spatial variable x, the format of the different basis vectors g_k are

$$g_k(x) = \cos(\pi k x), \tag{4.6}$$

where k represents the wave number (spatial frequency). Similarly when considering functions defined over time and not space, the basis vectors are defined as

$$g_\omega(t) = \cos(\pi \omega t), \tag{4.7}$$

where ω is the wave frequency. The alternative for odd functions is to develop them over a basis of sine waves (basis vectors for odd continuous functions) in a very similar way

$$g_k(x) = \sin(\pi k x), \tag{4.8}$$

where k represents the wave number (spatial frequency). Similarly when considering functions defined over time and not space, the basis vectors are defined as

$$g_\omega(t) = \sin(\pi \omega t). \tag{4.9}$$

With these vector bases, functions take the form

$$f(x) = \sum_{k=0}^{\infty} a_k g_k(x), \tag{4.10}$$

or

$$f(t) = \sum_{\omega=0}^{\infty} a_\omega g_\omega(x), \tag{4.11}$$

where the a_k or a_ω are coefficients that quantify the contribution of each basis vector into representing the function f. Finding these coefficients as well as other important properties of the basis vectors are discussed in Section 4.3.5. Section 4.3.4 presents a more general tool that applies to both odd and even functions (any continuous functions). It basically superposes the cosine and sine series and uses the fact that any continuous function can be built by a combination of odd and even functions.

4.3.4 **Fourier series representation**

In the section on complex calculus, we saw the Euler–Moivre formula (Eq. 2.22)

$$\exp(i\theta) = \cos(\theta) + i\sin(\theta). \tag{4.12}$$

In the present context, we are interested in representing any continuous function $f(x)$ as

$$f(x) = \sum_{k=0}^{\infty} a_k \cos(k\pi x) + b_k \sin(k\pi x). \tag{4.13}$$

One way to achieve this with the Euler–Moivre formula is to accept that the coefficients a_k, b_k may have an imaginary part (they are complex numbers) and then consider the real-valued function $f(x)$ as the real part of the sum of the waves

$$f(x) = Real\left(\sum_{k=0}^{\infty} a_k \cos(k\pi x) + ib_k \sin(k\pi x)\right) = Real\left(\sum_{k=0}^{\infty} c_k \exp(ik\pi x)\right). \tag{4.14}$$

This decomposition of the vectorial space of continuous function is generally referred to as Fourier series representation of functions. One of the very convenient and attractive features of cosine, sine, and Fourier series is that the derivative of the basis vector functions are very easy to compute. For the latter we find that

$$\frac{d\exp(ik\pi x)}{dx} = ik\pi \exp(ik\pi x). \tag{4.15}$$

4.3.5 **Orthogonality of sine, cosine, and exp basis functions**

The previous sections propose that sine, cosine, and exponential (Fourier) basis functions are appropriate choices for basis vectors for the vectorial space of continuous functions (the vectorial space of the types of function that would be solutions to our differential equations). We have yet to prove that they form a basis and that the representation of the functions they provide is well suited for the type of work we want them to perform. There are two main properties that we want to establish now: firstly, that these basis functions form a linearly independent set, and secondly, that they span the whole space of continuous functions they are meant to represent (even functions for cosine, odd for sine, and all of them for Fourier series). We start here with linear independence. With more conventional vectors (e.g., a basis of three-dimensional spatial vectors), linear independence means that we cannot build one of the basis function vectors from a linear combination of the others. That means that, if we collect the basis vectors (or continuous functions) in g_k for $k = 0, \ldots, \infty$, there are **no** choices of coefficients a_k to build a linear combination that can generate another basis vector, that is,

$$g_k(x) \neq \sum_{k' \neq k} a'_k g'_k(x). \tag{4.16}$$

With more conventional vectors we can see that the scalar product between two vectors allowed us to compute how one vector projects onto the other one. Here we expand on this idea and define a scalar product between functions. The scalar product is an operation that takes two vectors living in the same space (i.e., same dimensions) and projects them onto each other, the output being a scalar

$$\vec{a} \cdot \vec{b} = \lambda. \tag{4.17}$$

If $\lambda = 0$, it means that the two vectors are orthogonal to each other, that is, they do not project onto each other. Practically, for a set of N-dimensional vectors, the scalar product is

$$\sum_{j=1}^{N} a_j b_j = \lambda. \tag{4.18}$$

For continuous functions $f(x)$ and $g(x)$, it is possible to create similarly a scalar product, where the finite sum gives way to a definite integral. Here for the periodic functions we are observing, the integral is taken over the longest period of the two periodic functions involved,

$$\langle fg \rangle = \int_P f(x)g(x)\mathrm{d}x, \tag{4.19}$$

where P indicates the period and the scalar product is just a scalar number (not a function). Vectorial bases are by definition comprised of linearly independent vectors. We can, in many instances, request even more than independence: orthogonality. Orthogonality of the basis vectors simplifies how we compute the coefficients a_k or a_ω.

Starting with the cosine functions, defining a shorthand notation for the scalar product between the basis vectors associated with frequencies k and k'

$$I_{k,k'} = \int_{-P/2}^{P/2} \cos\left(\frac{2\pi kx}{P}\right)\cos\left(\frac{2\pi k'x}{P}\right)\mathrm{d}x, \tag{4.20}$$

we can first use the even behavior of the cosine functions to get

$$I_{k,k'} = 2\int_{0}^{P/2} \cos\left(\frac{2\pi kx}{P}\right)\cos\left(\frac{2\pi k'x}{P}\right)\mathrm{d}x. \tag{4.21}$$

First, we consider the case where the two modes, k and k', are the same, and we get

$$I_{k,k} = 2\int_{0}^{P/2} \cos^2\left(\frac{2\pi kx}{P}\right)\mathrm{d}x. \tag{4.22}$$

Using the trigonometric identity

$$\cos\alpha\cos\beta = \frac{1}{2}\left[\cos(\alpha-\beta) - \cos(\alpha+\beta)\right], \tag{4.23}$$

and setting $\alpha = \beta = \frac{2\pi kx}{P}$, we get $\alpha - \beta = 0$ and $\alpha + \beta = \frac{4\pi kx}{P}$ so that

$$I_{k,k} = \frac{2}{2}\int_{0}^{P/2} \left[\cos(0) + \cos\left(\frac{4\pi kx}{P}\right)\right]\mathrm{d}x = \frac{P}{2} + \underbrace{\frac{P}{4\pi k}\sin\left(\frac{4\pi kx}{P}\right)\Bigg|_0^{P/2}}_{=0} = \frac{P}{2}. \tag{4.24}$$

Now if $k \neq k'$, using the same trigonometric identity, we get

$$I_{k,k'} = \int_{-P/2}^{P/2} \cos\left(\frac{2\pi(k-k')x}{P}\right) \mathrm{dx} + \int_{-P/2}^{P/2} \cos\left(\frac{2\pi(k+k')x}{P}\right) \mathrm{dx}. \tag{4.25}$$

Integrating each term leads to

$$I_{k,k'} = \frac{P}{2\pi(k-k')} \sin\left(\frac{2\pi(k-k')x}{P}\right)_0^{P/2} + \frac{P}{2\pi(k+k')} \sin\left(\frac{2\pi(k+k')x}{P}\right)_0^{P/2}, \tag{4.26}$$

where $k + k'$ and $k - k'$ are integers, which means that the sine functions evaluated at $x = 0$ and $x = P/2$ are 0. Therefore,

$$I_{k,k'} = 0 \qquad \text{if } \mathrm{k}' \neq \mathrm{k}. \tag{4.27}$$

This proves that the cosine functions of the cosine series are orthogonal to each other

$$I_{k,k'} = \begin{cases} \frac{P}{2} & \text{if } k = k', \\ 0 & \text{otherwise.} \end{cases} \tag{4.28}$$

The same overall procedure applies to the sine series. Now the scalar product $I_{k,k'}$ refers to

$$I_{k,k'} = \int_{-P/2}^{P/2} \sin\left(\frac{2\pi kx}{P}\right) \sin\left(\frac{2\pi k'x}{P}\right) \mathrm{dx}, \tag{4.29}$$

and remembering that sine are odd functions, we get exactly the same outcome

$$I_{k,k'} = \begin{cases} \frac{P}{2} & \text{if } k = k', \\ 0 & \text{otherwise.} \end{cases} \tag{4.30}$$

We found that these sine and cosine functions form an orthogonal vectorial basis for odd and even functions (space of infinite dimension), respectively. Bringing cosine and sine series back together, we can now easily show that the Fourier series

$$f(x) = \sum_{k=0}^{\infty} a_k \exp\left(i\frac{2\pi kx}{P}\right) \tag{4.31}$$

is such that the complex exponential functions generate an orthogonal vectorial basis for continuous functions.

The question now is how we construct the series

$$f(x) = \sum_{k=0}^{\infty} a_k g_k(x), \tag{4.32}$$

for $g_k(x)$ a sine or cosine or complex exponential function. In other words, how do we get the coefficients a_k? Again this is where orthogonality helps tremendously. Think of orthogonality as a filter, and out of the infinite number of coefficients a_k present in the sum, orthogonality offers us a pair of tweezers to pick the coefficient of any particular mode (wavenumber k). To illustrate this property, we consider a function $f(x)$ represented by the cosine series

$$f(x) = \sum_{k=0}^{\infty} a_k \cos\left(\frac{2\pi kx}{P}\right). \tag{4.33}$$

Multiplying both sides by a given cosine with wavenumber k' and integrating over a period (for wavenumber k), we get

$$\int_{-P/2}^{P/2} f(x) \cos\left(\frac{2\pi k' x}{P}\right) \mathrm{dx} = \sum_{k=0}^{\infty} a_k \int_{-P/2}^{P/2} \cos\left(\frac{2\pi kx}{P}\right) \cos\left(\frac{2\pi k' x}{P}\right) \mathrm{dx} \tag{4.34}$$

where using the orthogonality of the basis functions, all the terms in the infinite sum drop out except when $k = k'$, leaving us with

$$\int_{-P/2}^{P/2} f(x) \cos\left(\frac{2\pi k' x}{P}\right) \mathrm{dx} = \frac{P}{2} a_{k'}, \tag{4.35}$$

which allows us to get any coefficient k' according to

$$a_{k'} = \frac{2}{P} \int_{-P/2}^{P/2} f(x) \cos\left(\frac{2\pi k' x}{P}\right) \mathrm{dx}. \tag{4.36}$$

The same approach applies also to the sine and Fourier series.

Chapter 5

Ordinary differential equations (ODEs)

Ordinary differential equations (ODEs) express a mathematical balance between terms expressed as rates or spatial variations (gradients) and are ubiquitous when dealing with mass, momentum, or energy balance statements. In contrast with partial differential equations (PDEs), ODEs codify the relationship between a function f and its sole dependent variable x or t in the examples provided in this chapter, that is, the function admits only a single variable. Part II of this book is devoted to the solution of ODEs with applications to Earth and planetary sciences and involves geochemical box models, magma chamber dynamics, the motion of planetary bodies due to gravity, and motion of charged particles in an electromagnetic field. This chapter reviews the classification of ODEs and discusses some methods to find analytical solutions, which is not always possible. Some of the analytical solutions are used later in the book. While interested readers are encouraged to take a deeper dive into the analysis of ODEs through devoted specialized textbooks (of which there are many), this chapter is intended to bring readers up to speed with a quick overview of selected important concepts. We start with a series of examples, define some characteristics of ODEs (e.g., linearity and homogeneity) and finally discuss some methods to retrieve analytical solutions.

5.1 Examples

Most (not all) examples discussed in this chapter relate to topics approached later in the book. We first derive or provide context for the differential equations.

5.1.1 Nuclear decay

A detailed derivation of the nuclear decay is presented at the beginning of Part II of this book. In summary, the statistical independence of decay events, that is, the probability of a new decay event, is independent of the timing of the previous event, leads to the decay process being characterized by an exponential decrease in decays with time (observed experimentally). If $N(t)$ represents the number of radioactive atoms present in the substance at time t, or more precisely in the interval of time $[t - dt/2, t + dt/2]$, then we find (see Part II) that the change in the number of radioactive atoms through time follows when $dt \to 0$

$$\frac{dN}{dt} = -\lambda N, \tag{5.1}$$

where λ is the decay constant. This is the first and simpler ODE we discuss here.

Introduction to Numerical Modeling in Earth and Planetary Sciences, Christian Huber, Oxford University Press.
© Christian Huber (2025). DOI: 10.1093/oso/9780198802716.003.0006

5.1.2 Harmonic oscillator

Assuming a mass m attached to a massless spring with a spring constant k, in the absence of other forces (gravity, friction), Newton's second law tells us that inertia balances the force of the spring

$$m\frac{d^2x}{dt^2} = -k(x - x_0), \tag{5.2}$$

where x is the position of the spring and $x - x_0$ is the displacement from the position at rest x_0. Defining displacement as $\hat{x} = x - x_0$, given x_0 is a constant, we can recast Newton's second law into

$$\frac{d^2\hat{x}}{dt^2} = -\frac{k}{m}\hat{x}. \tag{5.3}$$

This is our second ODE example and it differs from the first one mostly because here the derivative is of second order.

5.1.3 Cosmogenic nuclides steady-state no erosion nor sedimentation

In the previous two examples the variable to integrate was time. In the next two examples, we consider the concentration of ^{10}Be in a soil at depth z caused by cosmogenic nuclide production and nuclear decay. The equation that describes the evolution of the concentration of radiogenic elements in the soil is a PDE and its derivation is discussed in details in Part III, when discussing applications of the advection equation. We first consider the case here where the depth is fixed to a datum z_0 and assume no deposition of sediments or erosion. In this particular case, the original PDE (concentration depends on variables z, t) reduces to an ODE (t only). The ODE in the absence of sedimentation or erosion, at a fixed depth z_0, reduces to

$$\frac{dC}{dt} = -\lambda C + J_0 \exp\left(-\mu z_0\right), \tag{5.4}$$

where the first term is identical to the nuclear decay example (decay) and the latter term expresses the production rate of ^{10}Be at that reference depth.

5.1.4 Cosmogenic nuclides steady-state with sedimentation

We now consider the same system but under different conditions, that of steady-state (no temporal evolution). We seek the profile of concentration $C(z)$ under sedimentation conditions, where the concentration of ^{10}Be follows

$$u\frac{dC}{dz} = -\lambda C + J_0 \exp\left(-\mu z\right). \tag{5.5}$$

In this expression, the velocity u is constant and refers to the sedimentation rate of the soil.

5.2 Definitions

The first important characteristic of a differential equation is whether or not it is homogeneous. **Homogeneity** here refers to differential equations for the function $N(t)$ or $C(t)$ where all the terms depend explicitly on the function itself. For instance, the first two examples are homogeneous ODEs, because all terms of the first equation depend directly on N and similarly all terms in example two depend explicitly on $\hat{x}$. In contrast, the two latter examples contain a production term of the form $J_0 \exp(-\mu z)$ that does not depend explicitly on the function of interest C. They are therefore both examples of nonhomogeneous differential equations.

The second important concept is linearity. A linear differential equation is one where the function that we integrate for N or C above are not found elevated to other powers (C^2 or N^{-1}, for example) nor inserted into nonlinear functions (e.g., trigonometric functions or logarithm/exponential). As an example, one can test the linearity of a differential equation such that f is a solution of

$$\alpha \frac{df}{dt} = \beta f(t). \tag{5.6}$$

Assume that the function $g(t)$ is also a solution to this differential equation, because unless we specify an initial condition $f(t = 0)$ there is an infinite amount of solution to this differential equation. So

$$\alpha \frac{dg}{dt} = \beta g(t). \tag{5.7}$$

If we sum these two functions to create a new one, $h(t) = f(t) + g(t)$, then we find that $h(t)$ is also a solution to this differential equation

$$\alpha \frac{d(f+g)}{dt} = \beta(f+g). \tag{5.8}$$

Let us consider a counter-example with a slightly different ODE

$$\alpha \frac{df}{dt} = \beta f^2(t). \tag{5.9}$$

Now using the same approach leads to

$$\alpha \frac{d(f+g)}{dt} = \beta(f+g)^2 = \beta(f^2 + g^2 + 2fg). \tag{5.10}$$

The last term in the parenthesis clearly shows that the sum of f and g is not a solution of this ODE, and this is because it is not a linear equation.

Finally, the order of a differential equation is set by the highest order of the derivative in the equation. The first example is first order, the second is second order, and the latter two are first order. The order of the differential equation also influences the number of initial or boundary conditions required to fully define the problem. This is discussed further in the resolution of these examples below.

5.2.1 **Exercises**

- Determine which of the following ODEs are linear:

1.
$$\frac{dC}{dt} = -k(C - C_0), \tag{5.11}$$

where C is a concentration, k the rate constant is independent of C, and C_0 is the equilibrium concentration.

2.
$$\frac{dT}{dt} = \alpha(T^4 - T_0^4), \tag{5.12}$$

where α is a constant, T is the temperature of an object, and T_0 is a reference temperature.

3.
$$\frac{dA}{dt} = a_1 \cos(\omega t) - \lambda A, \tag{5.13}$$

where λ, ω, and a_1 are constants and A represents a conserved quantity of interest.

- Determine the order of each of the following differential equations and whether they are homogeneous:

1.
$$N\frac{dN}{dt} = -\lambda N, \tag{5.14}$$

where λ is a constant (please be careful with this one!).

2.
$$N\frac{dN}{dt} = -\lambda + \gamma N, \tag{5.15}$$

where λ and γ are constants.

3.
$$\frac{dC}{dt} = \alpha C + \cos(\omega C t), \tag{5.16}$$

where α, ω are constant.

5.3 Analytical solutions

In this section we show, by example, a few different approaches to finding analytical solutions to ODEs. The solution method depends on the homogeneity and nature of the nonhomogeneous term in the equation.

5.3.1 Nuclear decay

The ODE is

$$\frac{dN}{dt} = -\lambda N. \tag{5.17}$$

It is a linear homogeneous first-order ODE, which is the easiest family of ODE to solve. We can use a simple separation of the variables here to solve the equation. The approach is to separate N and t in the equation to get one side of the equation that depends solely on one, and the other side of the equation to depend solely on the other

$$\frac{dN}{N} = -\lambda dt. \tag{5.18}$$

The left side depends only on N and the right side only on t. The next step is to integrate each side from time $t = 0$ (initial condition) to the active time t

$$\int_{N(0)}^{N(t)} \frac{dN}{N} = -\int_{0}^{t} \lambda dt. \tag{5.19}$$

This yields

$$ln\left(\frac{N(t)}{N(0)}\right) = -\lambda t. \tag{5.20}$$

Manipulation of the logarithm leads to the analytical solution for the nuclear decay

$$N(t) = N(0)\exp(-\lambda t). \tag{5.21}$$

5.3.2 **Harmonic oscillator**

The simple harmonic oscillator leads to the simple ODE

$$\frac{d^2\hat{x}}{dt^2} = -\underbrace{\frac{k}{m}}_{\omega^2}\hat{x}. \tag{5.22}$$

This equation can be solved analytically once two initial conditions are provided. The two initial conditions relate to the second-order derivative over time (two constants of integration). We set $\hat{x}(t = 0) = \hat{x}_0$ and $d\hat{x}/dt(t = 0) = 0$, that is, that the position of the mass is initially not at rest but its initial velocity is 0. Recasting the equation in term of angular frequency yields

$$\frac{d^2\hat{x}}{dt^2} = -\omega^2\hat{x}. \tag{5.23}$$

Here we use another approach to find a solution to this second-order differential equation. Remember that trigonometric functions like cos and sin, when differentiated twice, are equal to themselves multiplied by a negative constant, so they seem to fit well with the ODE here. This approach is generally called that of the ansatz (a German word that means assumption or informed guess). We therefore seek for a general solution of the kind

$$\hat{x} = A\cos(\omega t) + B\sin(\omega t). \tag{5.24}$$

First, we check that the ansatz is correct and plug this general solution into the differential equation

$$\frac{d\hat{x}}{dt} = -\omega A\sin(\omega t) + \omega B\cos(\omega t). \tag{5.25}$$

The second derivative therefore yields

$$\frac{d^2\hat{x}}{dt^2} = -\omega^2 A\cos(\omega t) - \omega^2 B\sin(\omega t) = -\omega^2\hat{x}. \tag{5.26}$$

This shows that our ansatz was well informed. Now we can use the two initial conditions to determine the constants A and B from the general solution. This is an important point as there are generally an infinite family of solutions to any differential equation, but the solution becomes unique when initial (or boundary) conditions are specified. Using the first initial condition $\hat{x}(t=0) = \hat{x}_0$, we get

$$\hat{x}_0 = A \tag{5.27}$$

because $\cos(0) = 1$ and $\sin(0) = 0$. The second initial condition will take care of B, as

$$\frac{d\hat{x}}{dt} = -\omega A\sin(\omega t) + \omega B\cos(\omega t) \tag{5.28}$$

yields, when $t = 0$,

$$B = 0. \tag{5.29}$$

Overall the analytical solution is

$$\hat{x}(t) = \hat{x}_0\cos(\omega t). \tag{5.30}$$

5.3.3 Time-dependent cosmogenic nuclides

The next ODE is the third example above that codifies the time evolution of the cosmogenic nuclide concentration in a soil at a fixed depth, assuming no sediment deposition or erosion. The ODE reads

$$\frac{dC}{dt} = -\lambda C + J_0\exp(-\mu z_0). \tag{5.31}$$

This equation is complemented by an initial condition $C(t=0) = C_0$. This differential equation is linear, first order, and nonhomogeneous. We use this ODE to highlight how we deal with nonhomogeneous ODEs. The idea is to consider first the homogeneous version, which can be easily solved with separation of the variables (like the nuclear decay above). The homogeneous reduction of our ODE leads to

$$\frac{dC_h}{dt} = -\lambda C_h, \tag{5.32}$$

where the subscript h is there to remind us that C_h is not exactly the function we want to solve for, but rather its homogeneous version. By analogy to the nuclear decay, we get directly that

$$C_h(t) = A\exp(-\lambda t), \tag{5.33}$$

where A will be determined from the initial condition. This will be our homogeneous solution. We now turn to a particular solution C_p, which together with the homogeneous combines to make up the analytical solution for this ODE. A simple yet effective way to get the particular solution is to assume a solution C_p that is independent of time, which leads to

$$\lambda C_p = J_0 \exp(-\mu z_0). \tag{5.34}$$

The particular solution then is

$$C_p = \frac{J_0}{\lambda} \exp(-\mu z_0). \tag{5.35}$$

Now the analytical solution is of the form

$$C(t) = C_h(t) + C_p = A\exp(-\lambda t) + \frac{J_0}{\lambda} \exp(-\mu z_0). \tag{5.36}$$

We can determine the constant of integration A from the initial condition $C(t = 0) = C_0$ and get

$$C_0 = A + \frac{J_0}{\lambda} \exp(-\mu z_0), \tag{5.37}$$

so that

$$A = C_0 - \frac{J_0}{\lambda} \exp(-\mu z_0). \tag{5.38}$$

So overall we have the analytical solution

$$C(t) = C_0 \exp(-\lambda t) + \frac{J_0}{\lambda} \exp(-\mu z_0)(1 - \exp(-\lambda t)). \tag{5.39}$$

5.3.4 Steady cosmogenic nuclides profile during sedimentation

The equation here is slightly different from the previous example as the primary variable is soil depth z and the equation is

$$u\frac{dC}{dz} = -\lambda C + J_0 \exp(-\mu z). \tag{5.40}$$

This equation needs to be supplemented with a boundary condition, here $C(z = 0) = C_0$. It is a first-order, linear, and nonhomogeneous ODE. However, the nonhomogeneous term depends on the primary variable, which does not allow us to use the same simple trick for the particular solution (C_p such that $dC_p/dz = 0$ would not work anymore). We therefore need to introduce a new method, generally referred to as the method of the variation of the constant. It also involves the homogeneous solution but the strategy is different beyond that point. The homogeneous solution is such that

$$u\frac{dC_h}{dz} = -\lambda C_h, \tag{5.41}$$

which can be solved with a separation of the variable and integrated to get

$$C_h(z) = A\exp\left(-\frac{\lambda z}{u}\right). \tag{5.42}$$

The factor A was generally referred to as a constant of integration and related to the initial or here boundary condition. We now consider that A is a function of the primary variable z in the general solution (not the homogeneous). This trick allows us to get rid of a term in the ODE. Now we assume the solution to the nonhomogeneous equation to be

$$C(z) = A(z)\exp\left(-\frac{\lambda z}{u}\right). \tag{5.43}$$

We insert this expression back into our original ODE using

$$\frac{dC}{dz} = \frac{dA}{dz}\exp\left(-\frac{\lambda z}{u}\right) - A\frac{\lambda}{u}\exp\left(-\frac{\lambda z}{u}\right) \tag{5.44}$$

to get

$$u\frac{dA}{dz}\exp\left(-\frac{\lambda z}{u}\right) - \cancel{A\lambda\exp\left(-\frac{\lambda z}{u}\right)} = \cancel{-A\lambda\exp\left(-\frac{\lambda z}{u}\right)} + J_0\exp\left(-\mu z\right). \tag{5.45}$$

This equation can be recast as

$$\frac{dA}{dz} = \frac{J_0}{u}\exp\left(-\left(\frac{\mu u - \lambda}{u}\right)z\right) \tag{5.46}$$

to solve for $A(z)$. Integrating, we get

$$A(z) = B - \frac{J_0}{\mu u - \lambda}\exp\left(-\left(\frac{\mu u - \lambda}{u}\right)z\right). \tag{5.47}$$

The general solution of the ODE is then

$$C(z) = A(z)\exp\left(-\frac{\lambda z}{u}\right) = \left[B - \frac{J_0}{\mu u - \lambda}\exp\left(-\left(\frac{\mu u - \lambda}{u}\right)z\right)\right]\exp\left(-\frac{\lambda z}{u}\right). \tag{5.48}$$

Using the boundary condition $C(z = 0) = C_0$ to define B yields

$$B = C_0 + \frac{J_0}{\mu u - \lambda}, \tag{5.49}$$

and the analytical solution is

$$C(z) = C_0 \exp\left(-\frac{\lambda z}{u}\right) + \frac{J_0}{\mu u - \lambda}\left[\exp\left(-\frac{\lambda z}{u}\right) - \exp\left(-\mu z\right)\right]. \tag{5.50}$$

This analytical expression is used in the Part III when discussing the application of the advection equation to cosmogenic nuclides.

Chapter 6

Partial differential equations (PDEs)

6.1 Eulerian versus Lagrangian reference frames, material derivatives

When considering problems that involve both time and space, and therefore functions (pressure, temperature, density, and concentration, to name a few scalar examples) that depend on both a time variable and position, it is possible and sometimes favorable to consider different options for reference frames. Specifically, there are circumstances where moving reference frames may offer some significant advantages mathematically. Let us consider friends hiking up from a valley to the top of a nearby mountain, where the elevation change is such that it causes measurable differences of temperature caused by the lapse rate in the atmosphere (temperatures get colder at higher elevation). The friends decide to use different means of transportation for their trip up the mountain: one (A) uses a helicopter; the second (B) uses a car driving up a road leading to the summit; and the third (C) goes on foot. They start at the same time (noon, $t = 0$ for us) and from the same point in the valley ($\vec{x} = 0$, center of our fixed reference frame). Assuming a fixed lapse rate for convenience (not dependent on time) $\alpha = \frac{\partial T}{\partial z} < 0$, we can reason that A travels quickly enough up to the summit that the temperature she measures during her travel to the top of the mountain mostly follows the lapse rate

$$T(\vec{x}, t) \approx T(0,0) + \alpha \frac{dz}{dt} t, \tag{6.1}$$

where $\frac{dz}{dt}$ is the average vertical component of the velocity of the helicopter over the path from the valley to the summit. In other words, the change in temperature experienced by A is dominated by the transport from a warmer place (valley) to a colder place (top of the mountain), rather than local changes in temperature because the trip is quick. On the other end of the spectrum, C's hike to the summit takes a long time (half a day, for example) and while the temperature overall is hot at the beginning (noon) it gets colder with time because of diurnal change of temperature. This effect is compounded to the effect of the lapse rate to lead to a greater amount of temperature loss along her journey up the mountain. Friend B experiences something in between. Let us write the rate of temperature change measured by A during her journey as $\frac{DT}{Dt}_A$ and use a similar notation for B and C. These measurements are made in reference frames that move at different paces across the same (or similar) landscape(s). After comparing their measurements it becomes clear that the rates are different and that the magnitude of the temperature change is also different. In order to understand these different measurements, it is important to reflect on what drives temperature changes for each traveler. In the case of A, the change of temperature over a short interval of time δt is δT

$$\delta T = \frac{\partial T}{\partial t}\delta t + \underbrace{\frac{\partial T}{\partial z}}_{\alpha} V_z^A \delta t, \tag{6.2}$$

Introduction to Numerical Modeling in Earth and Planetary Sciences, Christian Huber, Oxford University Press.
© Christian Huber (2025). DOI: 10.1093/oso/9780198802716.003.0007

where $\frac{\partial T}{\partial t}$ is the local (fixed in space) change in temperature over time and V_z^A is the vertical component of the velocity of A. If the travel is much faster than diurnal changes of temperature, the first term of the right-hand side of this equation may be negligible compared to the second term, which describes the changes in temperature caused by traveling in a heterogeneous temperature field. In summary, we learn that there are two contributions to δT:, one stemming from local (fixed in space) temporal changes in temperature (e.g., diurnal effect), and one associated with the movement of the temperature probe across a heterogeneous temperature field. For A, the latter is expected to dominate δT. In the case of C, the same formalism is true

$$\delta T = \frac{\partial T}{\partial t}\delta t + \underbrace{\frac{\partial T}{\partial z}}_{\alpha} V_z^C \delta t, \tag{6.3}$$

but the latter term is significantly smaller given V_z^C is small. As a result we expect the first term $\frac{\partial T}{\partial t}\delta t$ to play a more important role than for A. The case of B is likely to be intermediate between A and C.

What did we learn from this exercise? First, that in a moving reference frame there are two contributions to changes in temperature (or any other measurable field): a change that is purely temporal and local caused here by $\frac{\partial T}{\partial t}\delta t$; and that, if one moves in a heterogeneous field, a contribution $\frac{\partial T}{\partial z}V_z^A\delta t$ is added. Ultimately the rate of change in temperature can easily be retrieved by dividing both sides of Eq. 6.3 by δt and taking the limit $\delta t \to 0$ to get

$$\frac{DT}{Dt} = \frac{\partial T}{\partial t} + \frac{\partial T}{\partial z}V_z, \tag{6.4}$$

in one dimension, or more generally

$$\frac{DT}{Dt} = \frac{\partial T}{\partial t} + \vec{V} \cdot \nabla T. \tag{6.5}$$

We call the fixed reference frame where temperature change is purely equal to $\frac{\partial T}{\partial t}$ a **Eulerian reference frame**, while the moving reference frame is generally considered a **Lagrangian reference frame** when it moves according to the velocity field of the moving object. We note that the rate of temperature change in the two reference frames is different because of $\vec{V} \cdot \nabla T$ when we move across a heterogeneous temperature field.

6.2 Classification of second-order PDEs

Second-order PDEs are those that do not include terms with derivatives of a higher order than second. In most fields of mechanics, conservation laws are generally expressed as first- or second-order PDEs and therefore the classification for second-order PDEs presented here, which includes first-order PDEs as well, is appropriate for most of the practical examples in this book. First let us define the most generic second-order PDE (for two variables only) as

$$A\frac{\partial^2 T}{\partial x_1^2} + B\frac{\partial^2 T}{\partial x_1 \partial x_2} + C\frac{\partial^2 T}{\partial x_2^2} + D\frac{\partial T}{\partial x_1} + E\frac{\partial T}{\partial x_2} + FT = G, \tag{6.6}$$

where A, B, C, D, E, F, and G are constants or functions of x_1, x_2. First, the PDEs are called homogeneous if $G = 0$ for all choices of x_1, x_2.

There are three fundamental types of PDEs here: **parabolic**, **hyperbolic**, and **elliptic** equations.

6.2.1 Hyperbolic equations

Hyperbolic equations are defined by $B^2 - 4\,AC > 0$. They commonly involve time and space and are prone to develop wave-like behavior. Admittedly the poster child of hyperbolic PDEs is the wave equation (here for second-order PDEs):

$$\frac{\partial^2 T}{\partial t^2} = c^2 \frac{\partial^2 T}{\partial x^2}. \tag{6.7}$$

A quick use of the definition shows that $B^2 - 4\,AC = 4c^2 > 0$, where c (distance/time) is the wave speed.

6.2.2 Parabolic equations

Parabolic PDEs are such that $B^2 - 4\,AC = 0$. Generally the variables involve both space and time. The most common example of a parabolic PDE is the diffusion equation

$$\frac{\partial T}{\partial t} = D \frac{\partial^2 T}{\partial x^2}, \tag{6.8}$$

where D (distance^2/time) is a diffusion coefficient. A quick inspection of the differential equation shows that $B = 0$ and one of A, C is also 0.

6.2.3 Elliptic equations

Elliptic equations satisfy the condition $B^2 - 4AC < 0$. Laplace's and Poisson's equations

$$\frac{\partial^2 T}{\partial x_1^2} + \frac{\partial^2 T}{\partial x_2^2} = 0, \qquad \text{Laplace} \tag{6.9}$$

$$\frac{\partial^2 T}{\partial x_1^2} + \frac{\partial^2 T}{\partial x_2^2} = \rho(x_1, x_2), \qquad \text{Poisson} \tag{6.10}$$

are respectively homogeneous and nonhomogeneous second-order elliptic PDEs. Elliptic equations differ in an important way from hyperbolic and parabolic equations in that they do not admit characteristic curves, that is, curves (in time and space) along which information travels.

6.2.4 Method of characteristics for certain PDEs

The method of characteristics is a powerful method to solve some PDEs (hyperbolic and parabolic). With this method, the strategy is to find characteristic curves (parameterized path or surfaces in the spatial and temporal variables) that are such that if one casts a hyperbolic or parabolic PDE in terms of characteristic curves, these equations reduce to ODEs that can be solved with conventional ODE solution procedures. In Part III we look at one case where the method is applied to the linear advection equation. Characteristic curves can also be useful when selecting a suitable discretization for finite difference equations that ensures stability of the numerical scheme. We encourage the interested reader to look for more advanced texts in the theory of PDEs to learn more about the method of characteristics and the derivation of characteristic surfaces. The point we make here is rather qualitative: characteristic surfaces are defined by functions $f(x_1, x_2) = 0$, for example (equations of surfaces), such that when introduced into the PDE they yield, at any point, a null value

$$A\frac{\partial^2 f}{\partial x_1^2} + B\frac{\partial^2 f}{\partial x_1 \partial x_2} + C\frac{\partial^2 f}{\partial x_2^2} + D\frac{\partial f}{\partial x_1} + E\frac{\partial f}{\partial x_2} + Ff = 0. \tag{6.11}$$

In other words, the function f does not change along the characteristic surface. Because of this added constraint, if a characteristic curve is found, the PDE can be reduced to an ODE on these characteristic surfaces, allowing for simpler solution methods. Another way to frame these characteristic surfaces or curves is to see them as trajectories along which the information propagates. For instance, in a first-order linear advection equation (hyperbolic equation) characteristic curves are set by the transport coefficient (velocity field) and relate position and time through a simple integration. These curves reveal how the information (here, advection of any scalar field) propagates in the domain (in time and space). Interestingly characteristic curves and surfaces only exist for hyperbolic and parabolic PDEs, where they highlight how information is traveling and transmitted in a spatial domain when the system evolves transiently. On the other hand, elliptic problems do not admit characteristic curves or surfaces. Conceptually it should come as no surprise that they are generally describing steady-state systems. Poisson or Laplace equations would be solutions to the parabolic diffusion equation under steady-state conditions, for example.

Part II

Numerical modeling, ordinary differential equations (ODEs)

Chapter 7

First-order ODE (time integration): The nuclear decay equation as a starting point

7.1 The nuclear decay equation

The discovery of radioactivity is fairly recent (late nineteenth century), but it has greatly impacted our daily lives, the political scene, and Earth sciences. Radioactivity is commonly used now to provide absolute chronology (in contrast to relative chronology) and determine timescales associated with cosmochemistry, Earth differentiation, tectonics (fission tracking), geomorphology (e.g., cosmogenic nuclides), and magmatic processes. Radioactivity was discovered by Henri Becquerel, who showed that particle emissions from uranium salts were able to leave a record on nearby photographic plates. There are different types of nuclear radiation (alpha, beta, and gamma) that can be related to the emission of different particles (or nucleus for alpha). For the sake of the discussion in this chapter we focus first on radioactive nuclides (parents) that are themselves not produced by another decay and that produce a stable nuclide (daughters). We start with the exponential nuclear decay law because it is a first-order ordinary differential equation (ODE) that admits a simple analytical solution (shown later on). Our goal here is to introduce the foundation of the finite difference method and discuss some fundamental issues associated with the approximations involved in the numerical procedure to solve differential equations.

Before introducing the mathematical framework to develop a finite difference equation (FDE) from an ODE, let us discuss the derivation of the nuclear decay law under two different perspectives: a phenomenological approach and a mathematical description of the underlying statistical process behind nuclear decay. The first approach considers that we have access to a radioactive sample as Becquerel did and a device that allows us to record the number of decays in the sample as function of time (e.g., a Geiger counter). We introduce the observable $A(t)$, the activity of the substance in the sample, which is the number of decays per unit time that we are recording. We can record how the activity of the sample varies over time and retrieve a decay similar to that observed in Figure 7.1. The analysis of this time series is easier when the activity is actually replaced graphically by the (natural) logarithm of $A(t)$. In this plot, we observe a linear decay of $\log(A)$ with time. The slope in this linear trend is a constant $-\lambda$

$$\frac{d(\log A)}{dt} = \frac{1}{A}\frac{dA}{dt} = -\lambda, \tag{7.1}$$

which has units 1/time. The slope represents the rate of activity decrease in the sample and λ will be called the decay constant. This differential equation can be solved easily with the method of the separation of variables (see Chapter 5)

$$\frac{dA}{A} = -\lambda dt, \tag{7.2}$$

Introduction to Numerical Modeling in Earth and Planetary Sciences, Christian Huber, Oxford University Press.
© Christian Huber (2025). DOI: 10.1093/oso/9780198802716.003.0008

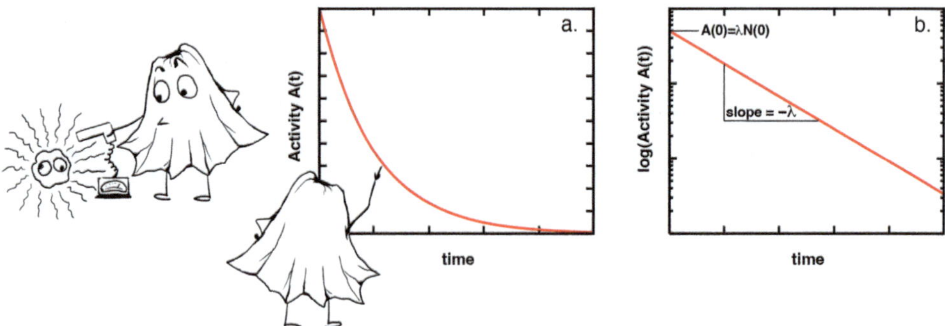

Figure 7.1 Activity measured on a radioactive sample versus time. Determining the decay constant (slope) and initial activity (intercept)

which yields

$$A(t) = A(0)\exp(-\lambda t), \tag{7.3}$$

where $A(0)$ is the activity of the sample at time $t = 0$, which is the intercept of the straight line in Figure 7.1. Now, using the definition of activity, that is, the number of decay events in the sample per unit time, $A \equiv -dN/dt$, where $N(t)$ is the number of radionuclides in the sample at time t. We can then conclude that

$$\frac{dN}{dt} = -A(0)\exp(-\lambda t), \tag{7.4}$$

which upon time integration yields

$$N(t) = \frac{A(0)}{\lambda}\exp(-\lambda t). \tag{7.5}$$

At time $t = 0$ the number of radionuclides in the sample is therefore

$$N(0) = \frac{A(0)}{\lambda}, \tag{7.6}$$

and we retrieve the familiar nuclear decay equation

$$\frac{dN}{dt} = -\lambda N(= -A), \tag{7.7}$$

with solution

$$N(t) = N(0)\exp(-\lambda t). \tag{7.8}$$

The second approach to derive the nuclear decay equation is rooted in statistics. The decay of one atom in the sample is independent of the fate of other atoms. In other words, the occurrence of one or multiple decays over any given nonoverlapping interval of time are independent of each other, and the process follows therefore a nonhomogeneous Poisson statistical law. The differential probability for one atom to decay (event) over an infinitesimal time interval dt is equal to the product of the decay constant and the time interval dt

$$dP_1 = -\lambda dt, \tag{7.9}$$

please note here that this is a *differential* probability and therefore it can be negative. Let us now define the complementary probability that there is no event over an interval of time dt centered around t with $P_0(t) = 1 - P_1(t)$. Assuming that the atom considered did not decay during the first interval of time, the probability for that atom to remain intact over the next interval is the product of the probability of no events at both t and $t + dt$

$$P_0(t + dt) = P_0(t) \times (1 - \lambda dt). \tag{7.10}$$

Reorganizing this equation, we obtain

$$\frac{P_0(t + dt) - P_0(t)}{dt} = -\lambda P_0(t), \tag{7.11}$$

which at the limit where $dt \to 0$ yields the differential equation

$$\frac{dP_0}{dt} = -\lambda P_0, \tag{7.12}$$

which admits the solution $P_0(t) = \exp(-\lambda t)$. If the decay constant is small (small decay rate), the probability of no decay is $P_0 \to 1$ as expected. Now if instead of considering a single radioactive nucleus, we turn to N nuclei, then we retrieve the nuclear decay law from (Eq. 7.12) with the solution given by (Eq. 7.8). We are actually interested in the probability of a decay $P_1 = 1 - P_0$ during an interval dt, which allows us to write the differential probability of a decay as

$$dP_1 = d(1 - P_0) = \lambda \exp(-\lambda t)dt. \tag{7.13}$$

The time for the probability of decay to reach the value 1/2, also known as the half-life $t_{1/2}$ of the radionuclide, can be computed from the probability distribution (Eq. 7.13).

$$\frac{1}{2} = \int_0^{t_{1/2}} dP_1 = 1 - \exp(-\lambda t_{1/2}). \tag{7.14}$$

After reorganization, we retrieve the definition of the half-life

$$t_{1/2} = \frac{\ln(2)}{\lambda}. \tag{7.15}$$

Back to the ODE for nuclear decay

$$\frac{dN}{dt} = -\lambda N. \tag{7.16}$$

This ODE is linear; to prove it let $N(t) = \alpha N_1(t) + \beta N_2(t)$, with α and β real-valued constants

$$\frac{dN}{dt} = \frac{d(\alpha N_1 + \beta N_2)}{dt} = \alpha\frac{dN_1}{dt} + \beta\frac{dN_2}{dt}. \tag{7.17}$$

The right-hand side of the equation in turn becomes

$$\lambda N = \lambda(\alpha N_1 + \beta N_2), \tag{7.18}$$

and, reorganizing (7.17) and (7.18), we finally obtain

$$\alpha\left(\frac{dN_1}{dt} + \lambda N_1\right) = \beta\left(\frac{dN_2}{dt} + \lambda N_2\right), \tag{7.19}$$

where both parentheses are null because of (7.16). This proves the linearity of the ODE. It is homogeneous because $f(t, N)$ in

$$\frac{dN}{dt} = f(t, N) \tag{7.20}$$

does not include a term that does not depend on N, such as a constant or a function of time alone.

Obviously, the existence of an analytical solution does void the need for developing a numerical model to solve for the nuclear decay equation (7.16). With the knowledge of the initial condition, $N(0)$, the decay constant λ and the time t, the number of radioactive nuclei remaining in the sample can be calculated exactly with (7.8). We nevertheless use this simple example to introduce the approximation of ODEs using FDEs.

7.2 Numerical versus analytical solutions

Although this book is about learning how to build satisfying numerical approximations to solutions of differential equations that arise in science, we stress that analytical solutions should always be preferred to numerical approximations. If the problem that you need to study admits an analytical solution, you should **always** seek the analytical solution. After all, we show that the numerical solutions we get from FDEs are at best approximations with a limited accuracy. Accuracy is not the only reason to favor analytical solutions over numerical solutions, as in many cases estimation of the solution within small relative errors is quite acceptable. The real limitation of numerical approximations is that they provide no information about the structure of the solution. In other words, every time you are looking for an accurate solution to the differential equation but need to consider slightly different starting conditions, you have no choice but to run your model for each specific case.

In order to illustrate the difference between an analytical and a numerical solution, let us assume that you visit a country and that you have no knowledge of the local geography. If you are left without a map or any means to get one (no internet) and cannot read the road signs, each trip requires asking locals for directions. As we know from experience, asking for directions can be considered a random process, where the quality of the information depends on the knowledge of the person helping you and the clarity of their communication. These are, to some extent, factors that you do not control (equivalent to round-off errors, but with probably bigger implications! See Section 7.8.2). Unless the next trip that you need to do is identical to a previous one, you must ask for directions each time. If, however, you prepare for your next travel by purchasing maps with the appropriate level of details, you can bypass the external help and prepare any itinerary just by checking your map. The maps in this case are your analytical solution; they contain all the information required, that is, they tell you how to react to any different choice of travel. On the other hand, numerical solutions provide no sense of the local geography of the place you visit, you only learn about the geography of a given point in the map by visiting it and need external help to get there (where some error can easily build up).

So if analytical solutions are so much better than numerical solutions, why do we bother discussing how to properly setup numerical solutions to differential equations? The sad answer is that most problems of scientific relevance are too complicated for us to find analytical solutions, so we are left with the next best approach: numerical approximations.

7.3 Using a Taylor series to discretize differential equations, the FDE

In order to solve differential equations, we need a way to treat differential operators (here dN/dt). Derivatives contain local info about the structure of the function N(t) around the point of interest. The definition of the first-order derivative of $N(t)$ is

$$\frac{dN}{dt} = \lim_{h\to 0} \frac{N(t+h) - N(t)}{h}, \tag{7.21}$$

where we again identify $h = dt$ to be an infinitesimal time interval. In what follows, dt is often referred to as the time step. In words, the rate of change of the number of radioactive nuclei in the sample is the limit of the difference obtained from measurements taken over an interval dt normalized by dt as the interval becomes infinitesimal. If this limit exists for any choice of $t\,(> 0)$, then $N(t)$ is differentiable (continuous) over that interval and the derivative is well defined. The idea behind the FDE is to replace derivatives with an approximation of its value by looking at the difference between $N(t+dt)$ and $N(t)$ over a finite but small interval dt. Because the interval is finite, the FDE is not exact and we can already guess that the accuracy of the approximation depends on the choice of interval dt. We start from a Taylor series of the function $N(t)$ defined by (7.16) to relate $N(t+dt)$ to $N(t)$ and derivatives of N at time t

$$N(t+dt) = N(t) + \frac{dN}{dt}|_t dt + \frac{1}{2}\frac{d^2N}{dt^2}|_t dt^2 + \frac{1}{6}\frac{d^3N}{dt^3}|_t dt^3 + \ldots + \frac{1}{n!}\frac{d^nN}{dt^n}|_t dt^n + \ldots, \tag{7.22}$$

where the symbol $|_t$ means that the derivative is estimated at time t. Subtracting $N(t)$ and then dividing both sides by dt, we get

$$\frac{N(t+dt) - N(t)}{dt} = \frac{dN}{dt}|_t + \frac{1}{2}\frac{d^2N}{dt^2}|_t dt + \frac{1}{6}\frac{d^3N}{dt^3}|_t dt^2 + \cdots + \frac{1}{n!}\frac{d^nN}{dt^n}|_t dt^{n-1} + \ldots. \tag{7.23}$$

In the limit of small time interval $dt \ll 1$, we can write

$$\frac{N(t+dt) - N(t)}{dt} = \frac{dN}{dt}|_t + \mathcal{O}(dt) \underbrace{\approx}_{dt\ll 1} \frac{dN}{dt}|_t, \tag{7.24}$$

where the latter step is just an approximation where dt is assumed sufficiently small so that we can truncate the Taylor series and neglect all terms with order $\geq dt$. The terms left out in (Eq. 20.24)

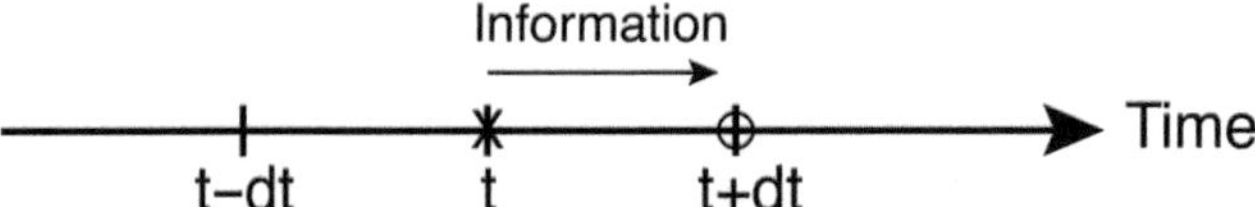

Figure 7.2 Stencil for the forward Euler model (same for backward Euler). The open circle symbol stands for the unknown that we are approximating and the cross represents the previous iteration (known value). The information follows time and skips one step at a time, and there is only one piece of information required ($N(t)$) to compute the next solution $N(t + dt)$.

constitute what we refer to as the *truncation error*. We obtain an approximation of the ODE that contains the variation (difference) of the function N over the finite interval dt in place of the derivative. This is the FDE

$$\frac{N(t + dt) - N(t)}{dt} = -\lambda N. \tag{7.25}$$

Because the FDE is only an approximation of the original ODE (and is exact only when $dt \to 0$), the solution of the FDE, in most cases, will not be equal to that of the ODE. We hope that the approximation is good enough so that the FDE solution approaches (converges) to that of the ODE when the choice of the finite time interval dt is carefully selected. This is the main assumption and challenge behind finite difference modeling. There are obviously different ways to approximate derivatives with finite differences, and we next review some classic examples.

7.4 Forward Euler (FE) discretization

We first note that we purposely left out any notion of time on the right-hand side of (Eq. 7.25). The first choice, and maybe the more instinctive, is to assume that

$$\frac{N(t + dt) - N(t)}{dt} = -\lambda N(t), \tag{7.26}$$

and that we construct an estimate for the number of radioactive atoms at time $t + dt$ only from the knowledge of the number of atoms at the previous time t, the decay constant, and the choice of time interval dt. Diagrams that show how the information propagates from a time step to the next are called stencils and the example for a first-order forward Euler (FE) method (e.g., nuclear decay) is shown in Figure 7.2.

Time integration requires an initial condition as a starting point. Once a starting value is provided, $N(0)$, one can see that $N(dt)$ is the only unknown quantity and can be estimated with the FDE. Then, the next step can be computed similarly to get $N(2dt)$ from $N(dt)$. This process is referred to as a marching algorithm. In order to better highlight this property of the FDE, let us first reorganize it

$$N(t + dt) = N(t)\,(1 - \lambda dt)\,. \tag{7.27}$$

Figure 7.3 Inputs outputs.

Starting from the initial condition $N(0)$, we obtain

$$\begin{aligned} N(dt) &= N(0)\,(1-\lambda dt) \\ N(2dt) &= N(dt)\,(1-\lambda dt) = N(0)\,(1-\lambda dt)^2 \\ \ldots &= \ldots \\ N(kdt) &= N(0)\,(1-\lambda dt)^k . \end{aligned} \tag{7.28}$$

In terms of implementation, the following pseudo-code highlights the simplicity of the FE scheme for this particular ODE

```
N(1)=N0;                        % Initial condition
for time=1:maxtime-1                        % Marching equation
    N(time+1)=N(time)*(1-lambda*dt);  % Forward Euler FDE
end
```

The advantage of this simple example is that we have the analytical solution to the nuclear decay law and hence we can assess the quality of the numerical (approximated) solution. We expect the accuracy of the approximation to depend on the time step, dt, but we also want to see if the value of the decay constant influences the numerical solution. Table 7.1 shows snapshots that represent the evolution of the number of radioactive atoms left in the sample up to $t = 20$; here the time units are not important for the sake of the discussion.

At a first glance we observe that the behavior of the solution (fit to the analytical solution) is much better when the decay constant λ and the time step dt are both small, that is, here when $dt = 1$ and $\lambda = 0.1$. As either dt or λ increase, the accuracy of the model decreases and the result becomes questionable at best until the numerical solution oscillates and its amplitude grows unbounded when $dt = 2$ and $\lambda = 1.2$. The behavior of the solution is explained in greater detail in Section 7.10. In the meantime, the take-away message from this section is that the quality of the numerical approximation with the FE method for nuclear decay depends strongly on the choice of time step (and to some extent on the decay constant that characterizes the radioactive process).

The FE approximation for the time derivative is not the only possible choice. It is important therefore to contrast these results with other approximations.

Table 7.1 Forward Euler nuclear decay calculations for different sets of time intervals dt and decay constants λ. The total time here maxtime × dt = 20 for all calculations.

	$\lambda = 0.1$	$\lambda = 0.8$	$\lambda = 1.2$
$dt = 1$			
$dt = 2$			

7.5 **Backward Euler (BE) discretization**

Drawing from a similar approximation as the FE, we may wonder what would happen if the activity $-\lambda N(t)$ on the right-hand side of the FE FDE is replaced by $-\lambda N(t+dt)$. Repeating the same approximation as for the FE, we assume that dt is small enough to neglect all terms of order dt. The FDE for the nuclear decay becomes then

$$\frac{N(t+dt)-N(t)}{dt} = -\lambda N(t+dt), \tag{7.29}$$

where the unknown $N(t+dt)$ reconstructed from the previous approximation $N(t)$ occurs on both sides of the equality. Note that the main difference between BE and FE methods is that the right-hand side of the equation (the nonderivative part) is taken at the new time. Isolating the unknown $N(t+dt)$, we obtain

$$N(t+dt) = \frac{N(t)}{1+\lambda dt}. \tag{7.30}$$

The BE FDE can also be recast as a marching equation

$$\begin{aligned} N(dt) &= \frac{N(0)}{1+\lambda dt} \\ N(2dt) &= \frac{N(dt)}{(1+\lambda dt} = \frac{N(0)}{(1+\lambda dt)^2} \\ \ldots &= \ldots \\ N(kdt) &= \frac{N(0)}{(1+\lambda dt)^k}. \end{aligned} \tag{7.31}$$

The stencil for BE is the same as for FE, information from $N(t)$ is propagated to estimate $N(t + dt)$. Writing a pseudo-code with the BE method shows that there is no real cost, in this particular case, to replace the FE with a BE approximation.

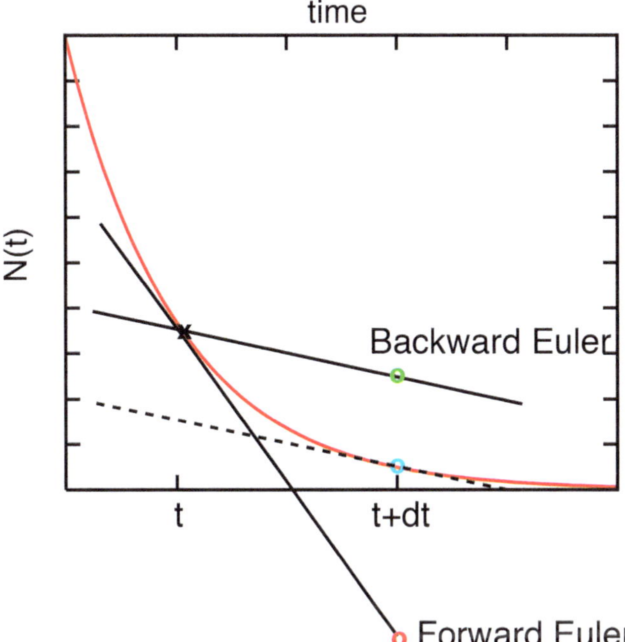

Figure 7.4 Difference between forward Euler and backward Euler approximations for the time derivative dN/dt. This assumes that we start the numerical integration from time $=t$ (initial condition). The forward Euler builds the new approximation $N(t + dt)$ from the slope at $N(t)$ (lower red circle), while the backward Euler method uses an approximation for dN/dt at $t + dt$ (dashed line). Using the slope at $t + dt$ the solution for time $t + dt$ is then projecting the same slope from the previous iteration (time $= t$) up to the green circle. The blue circle shows the analytical solution at $t + dt$.

```
N(1)=N0;                     % Initial condition
for time=1:maxtime-1                      % Marching equation
    N(time+1)=N(time)/(1+lambda*dt);  % Back. Euler FDE
end
```

Repeating the exact same calculations as for the FE model, we can compare the results from the BE model with the analytical solution (Table 7.2)

We find that the numerical approximation always behaves well, that is, it follows the analytical solution, but errors increase as λ and especially dt increase. Even though the FE and BE approximations are fairly similar for this particular example, the results are drastically different. For instance,

Table 7.2 Backward Euler nuclear decay calculations for different sets of time intervals dt and decay constants λ

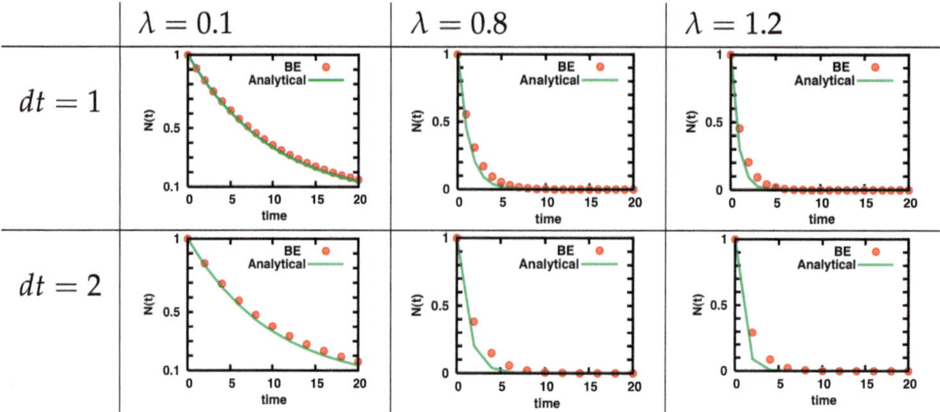

for large time steps and high decay constants, the FE model solution diverges with time while it remains bounded and follows the general trend expected from the analytical solution with the BE model approximation.

7.6 **Leapfrog or centered (CE) discretization**

The leapfrog (centered Euler, CE) method takes a different approach to the discretization of the time derivative. In essence, the idea is to retrieve a more accurate method than the former two and cancel naturally the bigger term left out during the truncation of the Taylor series ($\mathcal{O}(dt)$). Let us first use the Taylor series to construct $N(t + dt)$ and $N(t - dt)$ from $N(t)$

$$\begin{aligned} N(t+dt) = N(t) &+ \frac{dN}{dt}|_t dt + \frac{1}{2}\frac{d^2N}{dt^2}|_t dt^2 + \frac{1}{6}\frac{d^3N}{dt^3}|_t dt^3 \\ &+ \ldots + \frac{1}{n!}\frac{d^nN}{dt^n}|_t dt^n + \ldots, \end{aligned} \tag{7.32}$$

$$\begin{aligned} N(t-dt) = N(t) &- \frac{dN}{dt}|_t dt + \frac{1}{2}\frac{d^2N}{dt^2}|_t dt^2 - \frac{1}{6}\frac{d^3N}{dt^3}|_t dt^3 \\ &+ \ldots + (-1)^n\frac{1}{n!}\frac{d^nN}{dt^n}|_t dt^n + \ldots. \end{aligned} \tag{7.33}$$

Subtracting these two equations from one other, we get

$$\begin{aligned} N(t+dt) - N(t-dt) = \cancel{(N(t) - N(t))} &+ \left(\frac{dN}{dt}|_t dt + \frac{dN}{dt}|_t dt\right) + \cancel{\left(\frac{1}{2}\frac{d^2N}{dt^2}|_t dt^2 - \frac{1}{2}\frac{d^2N}{dt^2}|_t dt^2\right)} \\ &+ \left(\frac{1}{6}\frac{d^3N}{dt^3}|_t dt^3 + \frac{1}{6}\frac{d^3N}{dt^3}|_t dt^3\right) + \mathcal{O}(dt^4), \end{aligned} \tag{7.34}$$

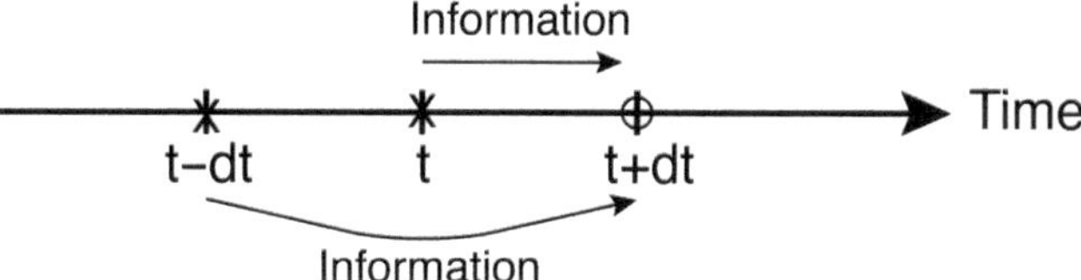

Figure 7.5 Stencil for the centered Euler or leapfrog discretization.

which gives us

$$\frac{N(t+dt)-N(t-dt)}{2dt} = \frac{dN}{dt}|_t + \mathcal{O}(dt^2). \tag{7.35}$$

Please note that because we cancel each even-order term in the series, the truncated terms depend on the time step as dt^2 now, instead of dt for the BE and FE. We therefore expect that the CE method yields more accurate results. Our FDE in this case becomes

$$\frac{N(t+dt)-N(t-dt)}{2dt} = -\lambda N(t), \tag{7.36}$$

and then the marching equation becomes

$$N(t+dt) = N(t-dt) - 2\lambda dtN(t). \tag{7.37}$$

The stencil for the CE approximation is shown in **Figure** 7.5 and illustrates why it is also commonly referred to as the leapfrog method. We note that this approximation requires N at both t and $t - dt$ to compute $N(t + dt)$. The improved accuracy comes at the cost of a more complex setup for the system to march the equation forward in time from the initial condition $N(0)$. Typically, we resort to computing the first time step $N(dt)$ with either BE or FE and then use the CE approximation for all subsequent steps. This is summarized in the following pseudo-code

```
N(1)=N0;                      % Initial condition
N(2)=N(1)/(1+lambda*dt);  %  First step using BE
for time=2:maxtime-1                    % Marching equation
    N(time+1)=N(time-1)-2*lambda*dt*N(time);  % CE FDE
end
```

The results we obtain are significantly different from what we observed with the FE and BE methods. First of all, we observe that most of the results display oscillations that grow with time and ultimately diverge, even over the short total time that we tested here (maxtime $\times$ dt = 20). Interestingly, it seems that the runs with the smaller chioce of time step and lower decay constant yield results that are consistent with the analytical solution. If we are to run the model with $\lambda = 0.1$ and $dt = 1$ for 100 time steps (maxtime = 100), we obtain the results shown in **Figure** 7.6. We find that none of the calculations presented here provides satisfying results with the CE approach for this particular problem (nuclear decay), that is, given enough time steps the approximated numerical solution diverges.

Table 7.3 Centered Euler nuclear decay calculations for different sets of time intervals dt and decay constants λ.

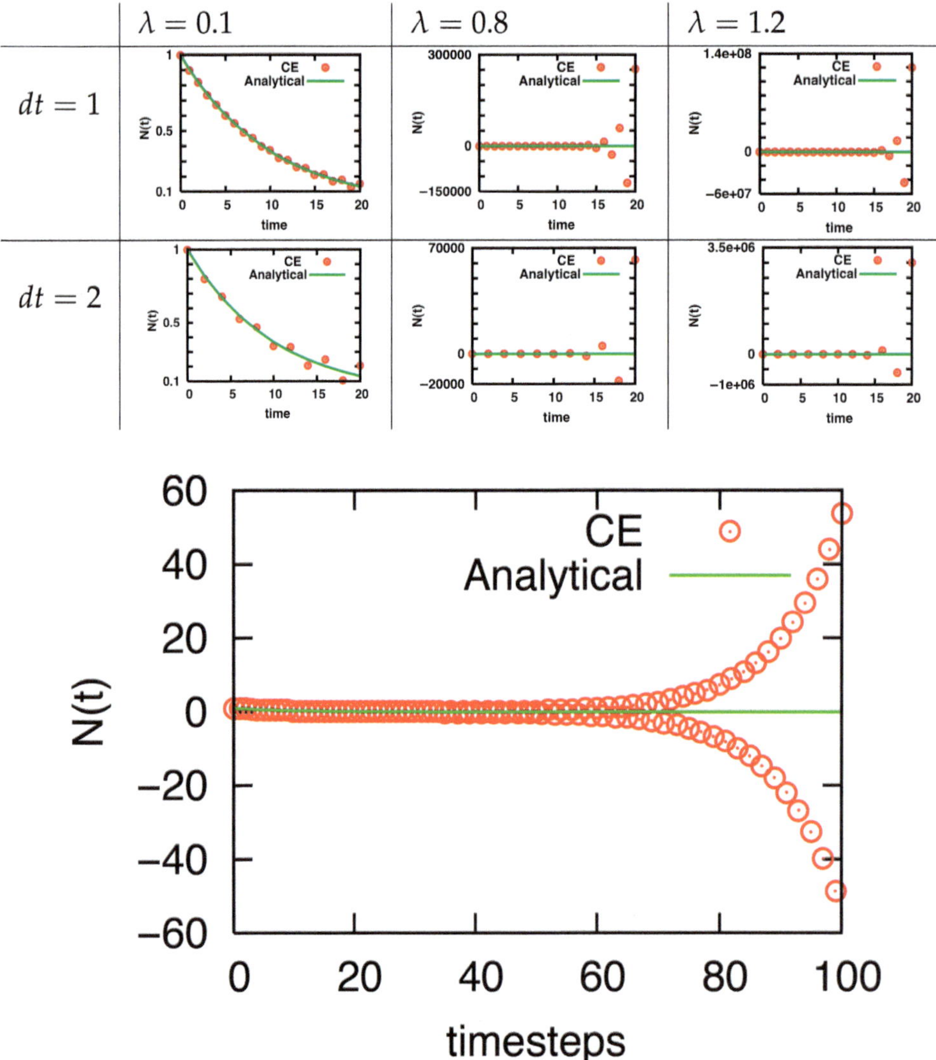

Figure 7.6 Results with the centered Euler approximation for $dt = 1$ and $\lambda = 0.1$ after 100 time steps. Oscillations become obvious after a few tens of time steps and the solution diverges quickly.

7.7 The order of a scheme

So far we have discussed three approaches for the discretization of the time derivative dN/dt. The first two models (FE and BE) were retrieved from a direct reorganization of the Taylor series and the omission of all terms that have a linear (and above) dependence on dt. These two methods involved only two points in the stencil $N(t)$ (known) and the next approximation $N(t + dt)$. The last approach was based on the comparison between two Taylor series expansions and offered an approximation for dN/dt where the lowest power of dt neglected was quadratic. The CE scheme however requires an extra point in the stencil, that is, $N(t+dt)$ is constructed explicitly from $N(t)$ and $N(t-dt)$.

The order of a scheme refers to the lowest power of the timestep dt for time derivatives (i.e., would be related to grid spacing dx for spatial derivatives, discussed later) that is truncated or discarded from our Taylor series expansion. For that reason, the FE and BE schemes presented here are both first order, while the CE method is second order. We also note that the number of points involved in the stencil of a scheme seems correlated to its order, and that higher-order schemes require more points (information) to build the next approximation.

The simplest example on how to build a higher-order scheme is illustrated in Section 7.6 with the CE method. We now want to generalize the approach for any order. We introduce real-valued coefficients a_j with $-k \leq j \leq k$ and multiply them to the Taylor series expansion for different times

$$a_k N(t+kdt) = a_k\left(N(t) + \frac{dN}{dt}|_t kdt + \frac{1}{2}\frac{d^2N}{dt^2}|_t (kdt)^2 + \frac{1}{6}\frac{d^3N}{dt^3}|_t (kdt)^3 + \ldots\right)$$

$$\ldots$$

$$a_1 N(t+dt) = a_1\left(N(t) + \frac{dN}{dt}|_t dt + \frac{1}{2}\frac{d^2N}{dt^2}|_t dt^2 + \frac{1}{6}\frac{d^3N}{dt^3}|_t dt^3 + \ldots\right)$$

$$a_{-1} N(t-dt) = a_{-1}\left(N(t) + \frac{dN}{dt}|_t (-dt) + \frac{1}{2}\frac{d^2N}{dt^2}|_t (-dt)^2 + \frac{1}{6}\frac{d^3N}{dt^3}|_t (-dt)^3 + \ldots\right)$$

$$\ldots$$

$$a_{-k} N(t-kdt) = a_{-k}\left(N(t) + \frac{dN}{dt}|_t (-kdt) + \frac{1}{2}\frac{d^2N}{dt^2}|_t (-kdt)^2 + \frac{1}{6}\frac{d^3N}{dt^3}|_t (-kdt)^3 + \ldots\right). \tag{7.38}$$

We now sum all of these equations together to get

$$\underbrace{\sum_{j=-k}^{k} a_j\left(N(t+jdt) - N(t)\right)}_{\text{stencil}} = \frac{\partial N}{\partial t}\underbrace{\sum_{j=-k}^{k} a_j(jdt)}_{A_1} + \frac{1}{2}\frac{\partial^2 N}{\partial t^2}\underbrace{\sum_{j=-k}^{k} a_j(jdt)^2}_{A_2}$$
$$\frac{1}{6}\frac{\partial^3 N}{\partial t^3}\underbrace{\sum_{j=-k}^{k} a_j(jdt)^3}_{A_3} + \ldots\,. \tag{7.39}$$

We collect the a_j into a vector $\mathbf{a}$ of size $2k$, and the A_i into a vector of size n (which depends on where the series is truncated). The objective now is to find the set of a_j or the vector $\mathbf{a}$ such that we can set the desired order with A_i. For example, using the CE scheme, we want $A_1 = 1$ and $A_2 = 0$, so that $n = 2$ using only $j = \pm 1$. Using the definitions of A_1 and A_2 we get

$$\sum_{j=-1,+1} a_j(jdt) = 1, \tag{7.40}$$

$$\sum_{j=-1,+1} a_j(jdt)^2 = 0. \tag{7.41}$$

These equations can be generally written in matrix form

$$\mathcal{M}_{ij}a_j = A_i, \tag{7.42}$$

where $\mathcal{M}_{ij} = (jdt)^i$, and for the CE we only consider $i = 1, 2$ and $j = -1, +1$. The goal now is to invert the matrix $\mathcal{M}$ to retrieve the coefficients a_j

$$a_j = \sum_i (\mathcal{M}_{ij})^{-1} A_i. \tag{7.43}$$

The first important point to make here is that for the inversion to be possible we need $\mathcal{M}$ to be a square matrix ($n = 2k$) and that its determinant is non null. It shows directly that imposing more constraints on the scheme (to cancel higher-order terms) increases n and therefore more points need to be considered in the stencil. Let us revert back to our CE scheme with $j = -1, +1$ and $i = 1, 2$. The matrix $\mathcal{M}$ and its inverse are

$$\mathcal{M} = \begin{pmatrix} -dt & dt \\ dt^2 & dt^2 \end{pmatrix}, \quad \mathcal{M}^{-1} = \frac{1}{2dt}\begin{pmatrix} -1 & 1/dt \\ 1 & 1/dt \end{pmatrix}, \tag{7.44}$$

which allows us to get

$$a_{-1} = -\frac{1}{2dt}, \tag{7.45}$$

$$a_1 = \frac{1}{2dt}, \tag{7.46}$$

which leads back to the FDE

$$\frac{N(t + dt) - N(t - dt)}{2dt} = -\lambda N(t). \tag{7.47}$$

The method is general and allows us to construct finite difference schemes of any order given that there are enough data points ($2k$) to satisfy the required constraints (n equations). There is another great upside to this general approach, which is that it allows us to construct finite difference equations with asymmetric stencils, which becomes important when considering boundary value problems (spatial derivatives instead of time derivatives). Assume we want to construct a second-order FDE for a spatial gradient but using an asymmetric scheme so that dU/dx at position x is constructed from $U(x)$, $U(x + dx)$, and $U(x + 2dx)$, for example. In this context, we could argue that the position x is located at a boundary of the domain and the information required to compute the gradient dU/dx at x is only available for positions $> x$. In this context, $j = 1, 2, A_1 = 1$, and $A_2 = 0$. The matrix $\mathcal{M}$ becomes

$$\mathcal{M} = \begin{pmatrix} dt & 2dt \\ dt^2 & 4dt^2 \end{pmatrix} \tag{7.48}$$

and its inverse is

$$\mathcal{M}^{-1} = \frac{1}{2dx}\begin{pmatrix} 4 & -2/dx \\ -1 & 1/dx \end{pmatrix}. \tag{7.49}$$

The coefficients are

$$a_1 = \frac{2}{dx}, \tag{7.50}$$

$$a_2 = \frac{-1}{2dx}, \tag{7.51}$$

which leads to the second-order FDE approximation

$$\begin{aligned} a_1 U(x+dx) &+ a_2 U(x+2dx) + (a_1 + a_2)U(x) \\ &= \frac{3U(x) + 4U(x+dx) - U(x+2dx)}{2dx} \approx \frac{dU}{dx}. \end{aligned} \tag{7.52}$$

7.8 Sources of error, truncation versus round-off

There are basically two sources of errors that come up while solving an ODE with a numerical approximation (FDE). As the first source of error is introduced by the user, it is therefore a *human* error. It comes from the substitution of finite differences in place of derivatives. As discussed, this error results from the truncation of the Taylor infinite series to a given order (in dt or dx, for example). This source of error in the numerical approximation is generally referred to as the *truncation error*. The second source of error is a computer-based error and comes from the fact that real numbers are encoded on finite memory spaces in computers. Memory requirement limits the number of bits that can be used to store a real number on a space of memory. Some numbers cannot be expressed exactly because of the memory limitations and that leads to the *round-off error*.

7.8.1 Truncation error

Truncation error is basically a problem *we* cause by replacing a differential equation (here so far ODE) with a Finite Difference approximation. We basically own responsibility for that error and need to find a way to control the damage it causes. This is the reason why we review multiple FDE schemes for the discretization of any differential equation, and why we want to assess how they compare and take the scheme that has ultimately the best compromise in terms of accuracy and efficiency (and obviously stability). The truncation error is associated, as its name suggests, with the truncation of the infinite Taylor series that allows us to replace derivatives (of any order) with a difference equation.

The FE method approximation was built in (Eq. 7.24) by neglecting all terms after the first derivative in the Taylor series

$$N(t+dt) = N(t) + \frac{dN}{dt}|_t dt + \underbrace{\frac{1}{2}\frac{d^2N}{dt^2}|_t dt^2 + \ldots + \frac{1}{n!}\frac{d^nN}{dt^n}|_t dt^n + \ldots}_{\text{neglected}}. \quad (7.53)$$

After reorganizing terms to extract the proper approximation for the first-time derivative, we get

$$\underbrace{\frac{N(t+dt)-N(t)}{dt} = \frac{dN}{dt}|_t}_{\text{FE approximation}} + \underbrace{\frac{1}{2}\frac{d^2N}{dt^2}|_t dt + \cdots + \frac{1}{n!}\frac{d^nN}{dt^n}|_t dt^{n-1} + \ldots}_{\text{neglected}}. \quad (7.54)$$

Using the FE approximation in lieu of the derivative, we retrieve the FDE

$$\underbrace{\frac{N(t+dt)-N(t)}{dt} = -\lambda N(t)}_{\text{FDE}} + \underbrace{\frac{1}{2}\frac{d^2N}{dt^2}|_t dt + \cdots + \frac{1}{n!}\frac{d^nN}{dt^n}|_t dt^{n-1} + \ldots}_{\text{Truncation error}}.$$

The truncation error for the FE approximation of the time derivative is led by the term

$$\frac{1}{2}\frac{d^2N}{dt^2}|_t dt, \quad (7.55)$$

which is first order (linear) with respect to our discretization (dt). This is a first-order scheme as seen previously. The behavior of all the terms neglected in the truncation error will partially control the behavior of the numerical scheme and its performance. At this stage it is important to distinguish two types of truncation errors that are often discussed in scientific publications. The first is the *local truncation error*, which is the error that builds up during a single time iteration, over dt. If we assume that $N(t)$ is exact, then the local truncation error is the amount of error that we get from the truncation at time $t + dt$. We can see that the truncation error always depends on a (positive) power of dt and therefore decreases as dt gets smaller. Another approach is to consider the *global truncation error*. It is important to note that the global truncation error is not the sum of the local truncation errors. In fact, the latter is defined as the truncation error over a single time step assuming that the solution at the previous time step is exact (see **Figure** 7.7).

For completeness, we can use a similar approach to derive the truncation error for the BE and contrast it with the FE discretization. Going back to the Taylor expansion in (Eq. 7.54) we get now

$$\underbrace{\frac{N(t+dt)-N(t)}{dt} = \frac{dN}{dt}|_t}_{\text{FE approximation}} + \underbrace{\frac{1}{2}\frac{d^2N}{dt^2}|_t dt + \ldots + \frac{1}{n!}\frac{d^nN}{dt^n}|_t dt^{n-1} + \ldots}_{\text{neglected}} = -\lambda N(t+dt). \quad (7.56)$$

For the latter equality, we use the definition of the BE FDE. Using again the Taylor expansion to recast $N(t+\Delta t)$ into a series around $N(t)$ yields

$$\frac{dN}{dt}|_t + \frac{1}{2}\frac{d^2N}{dt^2}|_t dt + \frac{dt^2}{6}\frac{d^3N}{dt^3}|_t + \ldots = -\lambda\left[N(t) + \frac{dN}{dt}|_t dt + \frac{1}{2}\frac{d^2N}{dt^2}|_t dt^2 + \ldots\right]. \quad (7.57)$$

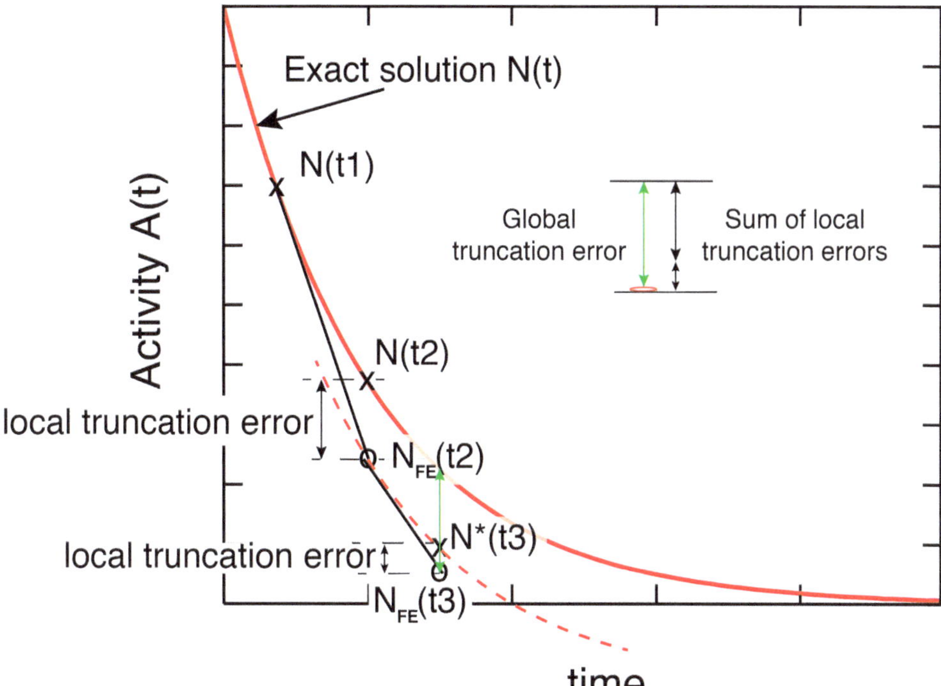

Figure 7.7 Global and local truncation errors. The analytical solution for the decay is shown for reference (thick red trend). The Forward Euler method is used to discretize the time derivative. We assume that we start from the exact number of atoms at time $t1$. $N_{FE}(t2)$ is then computed from the FDE and results in local truncation error ($N_{FE}(t2) \neq N(t2)$). To compute the local truncation error from $t2$ to $t3$, we consider that $N_{FE}(t2)$ is exact and start the analytical solution from there to retrieve the new "exact" solution $N^*(t3)$, which is now compared to $N_{FE}(t3)$. In contrast, the global truncation error between $t1$ and $t3$ is given by the green arrows. We see that the sum of the local truncation errors is not the same as the global error.

Reorganizing the equation, we get

$$\frac{dN}{dt}|_t = -\lambda N(t) \underbrace{-dt\left[\frac{1}{2}\frac{d^2N}{dt^2}|_t + \lambda\frac{dN}{dt}|_t\right] - \frac{dt^2}{2}\left[\lambda\frac{d^2N}{dt^2}|_t + \frac{1}{3}\frac{d^3N}{dt^3}|_t\right] + \ldots}_{\text{truncation error}} \quad (7.58)$$

The truncation error of the BE is slightly different from the FE model, that is, both depend on the first power of dt but the pre-factors are different, $-\left[\frac{1}{2}\frac{d^2N}{dt^2}|_t + \lambda\frac{dN}{dt}|_t\right]$ and $\frac{1}{2}\frac{d^2N}{dt^2}|_t dt$, for the BE and FE respectively.

7.8.2 **Round-off error**

The real numbers set is said to be dense, in other words, if a and b are any different ($a \neq b$) real numbers, then there is always a real number c in the interval between a and b. Then the same process can be repeated for the interval (a, c) or (c, b), and so forth. In between two different real values there is then an infinite number of real numbers. This property comes at a cost in terms of numerical calculations, because we can only represent a finite number of values with a finite memory space. For example, the storage of double floating point data types requires 64 bits of memory, which means that we can use up to 64 spaces that contain either a 0 or a 1 (binary numbers) to encode a number (that leads to 2^{64} possible combinations or different numbers). Out of these 64 bits, 1 is used to encode the sign of the number (±), 11 bits are used to encode the exponent, and the final 52 bits deal with what is often referred to as the significand or the mantisse

$$\underbrace{\pm}_{\text{sign}}\underbrace{\left(b_0 + b_1 2^{-1} + b_2 2^{-2} + \ldots b_{51} 2^{-51}\right)}_{\text{significand}}\underbrace{2^{c-1023}}_{\text{exponent}}, \tag{7.59}$$

where $1 \le c \le 2046$ and c is an integer. Although 64 bits offer a decent memory to encode most numbers, there are real numbers that cannot be represented exactly within this formalism. For example, $a = 0.1$ cannot be written exactly as a 64-bit floating point real number. This is a famous example and we encourage the reader to open a Matlab script and test

```
a=0.1;
b=0.3;
c=0.2;

% Compare a to (b-c)
% If the 2 expressions are equal to each other
% then we will get a 1 (true), 0 otherwise
a==b-c
>> 0 % the result you get from Matlab
```

So according to your computer, $0.3 - 0.2 \neq 0.1$, the problem is that 0.1 cannot be encoded exactly in floating point arithmetic. To see that better, if you now compute $b - c - a$ which you would hope goes to 0, you will obtain instead

```
>> b-a-c
ans=
  -2.7756e-17
```

Round-off errors are therefore related to our inability to write any possible (finite) real number in a given (fixed) allocated memory space. We refer to them as *computer errors* because there are limited ways we can avoid them, as opposed to *human errors*. Round-off errors can be limited to some extent by a careful approach to floating point arithmetics, but that rich topic is beyond the scope of this text. It is important to remember the danger of statements such as

```
>> a==b-c
```

because the outcome may turn wrong even though the algorithm is not essentially problematic!

7.9 **Convergence of a scheme**

Convergence is the holy grail of numerical integration of differential equations. A convergent scheme, given an appropriate choice of time step (and/or grid spacing), guarantees that the solution you obtain from the FDE approximation you are solving converges to the actual solution of the real ODE (or later PDE) you want to model. Convergence is very difficult to establish, but it is our goal throughout this textbook to test different algorithms for various differential equations and assess if they are problematic or if they converge to the desired solution. The challenge we face when assessing the convergence of a numerical scheme is that, in most cases, we do not know the analytical solution to which it should converge (otherwise the modeling effort is most likely pointless), so it is difficult to assess the performance of an FDE. Let us define the solution to a differential equation $N(t)$ (we assume for simplicity a time-dependent ODE) and the solution of the FDE approximation (FE, BE, CE, or any other scheme not yet discussed) $\mathcal{N}(t)$. Convergence means that

$$|N(t) - \mathcal{N}(t)| = \epsilon(dt) \rightarrow 0 \qquad \text{as } dt \rightarrow 0, \tag{7.60}$$

where $\epsilon(dt)$ is the error associated with our modeling efforts for the choice of time step dt. Convergence tells us that if we keep decreasing the time step, the approximation approaches the exact solution even closer. The rate at which the approximation converges to the exact solution as $dt \rightarrow 0$ depends on the order of the scheme. As discussed above, there are two contributions to the error ϵ: the round-off error and the truncation error that comes from replacing the ODE→FDE. We see shortly that there is a relatively straightforward approach to showing if a scheme is convergent; the strategy is to substitute the convergence condition (Eq. 7.60) with two conditions that are each easier to test using Lax equivalence theorem.

7.10 **Stability of FE, BE, and CE schemes**

The stability of a scheme refers to the conditions under which a numerical approximation (solution from our FDE) remains bounded for any number of iterations k given a time step dt. In mathematical terms, stability means that the solution N after any number of time steps k of duration dt is such that there exist a real positive finite constant C that satisfies

$$|N(kdt)| \leq C \qquad \text{for all } k. \tag{7.61}$$

The stability of a scheme depends on

1. The starting equation (ODE or PDE). Here we discuss the nuclear decay equation as an example. Note that a scheme that performs poorly for the nuclear decay may be excellent for another application.
2. The type and order of discretization scheme, that is, FE, BE, CE, or others presented later.
3. The choice of time step dt. There are three types of stability. *Unconditionally stable* stands for a scheme that is stable for any choice of dt for a given equation. *Unconditionally unstable* refers to a scheme that is always unstable (for any dt). A *conditionally stable* scheme is an FDE that is stable only over a range of possible time step values.

7.10.1 Stability of the FE model

Showing the stability of the FE model for a choice of time step dt for the nuclear decay equation amounts to making sure that $|N(kdt)| < C$ is true for any choice of k where C is a real (positive) constant. From the marching equation (Eq. 7.27), we can write the kth iteration for the estimation of the number of radioactive atoms left in the sample as function of its initial composition

$$N(kdt) = N(0)\,(1 - \lambda dt)^k. \tag{7.62}$$

Dividing both sides by $N(0)$ and considering the absolute value, we obtain the stability condition

$$\left|\frac{N(kdt)}{N(0)}\right| = |1 - \lambda dt|^k < \frac{C}{N(0)} \equiv C^*, \tag{7.63}$$

where it is easy to understand that $N(0) > 0$ for the case of nuclear decay. The numerical approximation after k iterations is hopefully positive, but because of truncation and round-off errors, it is possible for it to become negative (and, by extension worthless). The stability of the FE method for the nuclear decay equation depends on the behavior of $|1 - \lambda dt|^k$, that is, this expression must remain bounded for any choice of number of iterations k for the scheme to be stable for that choice of dt. As the number of iterations $k \to \infty$, stability then requires that $|1 - \lambda dt| \leq 1$. Any value of $|1 - \lambda dt|$ greater than 1 will diverge as k increases. The stability condition then reduces to

$$-1 \leq 1 - \lambda dt \leq 1, \tag{7.64}$$

which provides two conditions: that $\lambda \geq 0$ because dt is positive, and that $\lambda dt \leq 2$. For both conditions, equality yields a marginally stable condition (edge of the stable domain). Stability of the

Table 7.4 Forward Euler nuclear decay calculations for different sets of time intervals dt and decay constants λ. The total time here maxtime $\times$ dt = 20 for all calculations. The red cell shows the calculations that lay outside of the calculated stability field.

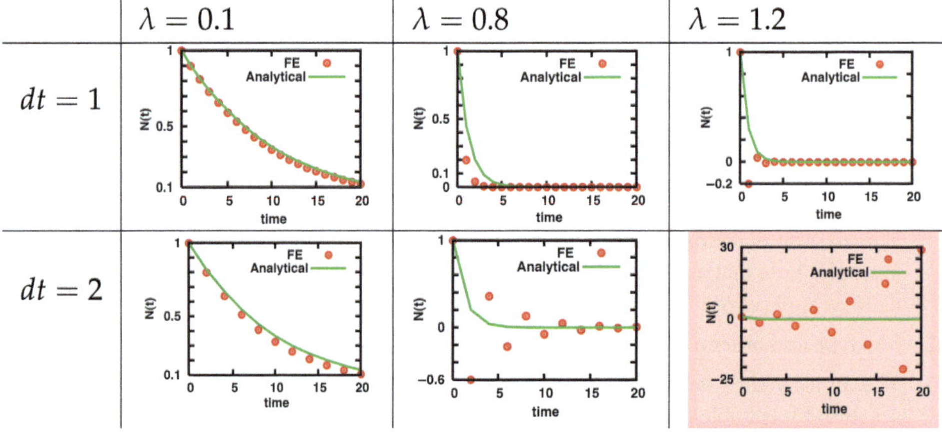

FE scheme for nuclear decay therefore depends on the choice of time step dt and decay constant λ, and this is called a *conditional stability*.

We are now ready to better understand the results shown in Table 7.3. The only clearly unstable simulation was that with $\lambda = 1.2$ and $dt = 2$, which violates the condition $\lambda dt \leq 2$. The run with $\lambda = 0.8$ and $dt = 2$ is stable but fairly inaccurate, especially for early times. The theoretical prediction for stability provides a valid way to assess the performance of the calculations presented in Table 7.4.

7.10.2 Stability of the BE model

We resort to a similar approach to demonstrate the conditions under which the BE scheme is stable for the nuclear decay equation. We use the FDE written as a marching equation (Eq. 7.31)

$$N(kdt) = \frac{N(0)}{(1 + \lambda dt)^k}. \tag{7.65}$$

A stable solution means that

$$\left| \frac{N(kdt)}{N(0)} \right| = \frac{1}{|1 + \lambda dt|^k} \leq C. \tag{7.66}$$

As k increases the denominator grows and the ratio decreases unless $|1 + \lambda dt| < 1$. Because $dt > 0$ the stability is guaranteed by $\lambda > 0$ because $1 + \lambda dt > 1$. The BE model is therefore unconditionally stable for the nuclear decay equation, that is, its stability does not depend on the choice of $dt > 0$. We observe from our calculations that the accuracy of the model decreases as the couple λdt increases (because of truncation errors) but that the solution remains bounded for every case we have tested, unlike the FE model.

7.10.3 Stability of the CE model

Let us first rewrite the FDE in the matrix form

$$\underbrace{\begin{pmatrix} N(kdt) \\ N((k+1)dt) \end{pmatrix}}_{\mathcal{N}_{k+1}} = \underbrace{\begin{pmatrix} 0 & 1 \\ 1 & -2\lambda dt \end{pmatrix}}_{\mathcal{M}} \underbrace{\begin{pmatrix} N((k-1)dt) \\ N(kdt) \end{pmatrix}}_{\mathcal{N}_k}. \tag{7.67}$$

In other words, we represent the following system of equations in matrix form (the first equation is trivial and the second describes our FDE):

$$\begin{aligned} N(kdt) &= N(kdt) \\ N((k+1)dt) &= N((k-1)dt) - 2\lambda dt N(kdt). \end{aligned}$$

This linear system can be used as a marching equation

$$\begin{aligned}\mathcal{N}_1 &= \mathcal{M}\mathcal{N}_0 \\ \mathcal{N}_2 &= \mathcal{M}\mathcal{N}_1 = \mathcal{M}^2\mathcal{N}_0 \\ &\dots \\ \mathcal{N}_k &= \mathcal{M}^k\mathcal{N}_0.\end{aligned} \tag{7.68}$$

The matrix $\mathcal{M}$ is real and symmetric so it can be diagonalized and the elements of the diagonal matrix will be the eigenvalues of $\mathcal{M}$ (see Section 3.4)

$$\mathcal{M} = \mathcal{Q}^{-1}\mathcal{D}\mathcal{Q}, \tag{7.69}$$

where $\mathcal{Q}$ is the transformation matrix (unitary matrix) and $\mathcal{D}$ is a diagonal matrix with eigenvalues of $\mathcal{M}$. Let us first see what $\mathcal{M}^2$ represents in terms of $\mathcal{Q}$ and $\mathcal{D}$[1]

$$\mathcal{M}^2 = \mathcal{Q}^{-1}\mathcal{D}\underbrace{\mathcal{Q}\mathcal{Q}^{-1}}_{\mathcal{I}}\mathcal{D}\mathcal{Q} = \mathcal{Q}^{-1}\mathcal{D}^2\mathcal{Q}, \tag{7.70}$$

and $\mathcal{D}^2$ is

$$\mathcal{D}^2 = \begin{pmatrix}\gamma_1 & 0 \\ 0 & \gamma_2\end{pmatrix}\begin{pmatrix}\gamma_1 & 0 \\ 0 & \gamma_2\end{pmatrix} = \begin{pmatrix}\gamma_1^2 & 0 \\ 0 & \gamma_2^2\end{pmatrix}, \tag{7.71}$$

with γ_1 and γ_2 the two eigenvalues of $\mathcal{M}$. The magnitude of the vector $\mathcal{N}_2$ is therefore

$$\mathcal{N}_2 = \mathcal{Q}^{-1}\begin{pmatrix}\gamma_1^2 & 0 \\ 0 & \gamma_2^2\end{pmatrix}\mathcal{Q}\mathcal{N}_0, \tag{7.72}$$

and because $\mathcal{Q}$ is a unitary matrix (so is its inverse), the amplitude of the solution $\mathcal{N}_2$ is only modified by a factor γ_1^2 and γ_2^2 from the initial condition $\mathcal{N}_0$. In order for the solution to our FDE to remain bounded, we need to make sure that the absolute value of both eigenvalues of $\mathcal{M}$, $|\gamma_i|$, remains smaller than 1. If that condition is not satisfied, after k iterations, the kth iteration is

$$\mathcal{N}_k = \mathcal{Q}^{-1}\underbrace{\begin{pmatrix}\gamma_1^k & 0 \\ 0 & \gamma_2^k\end{pmatrix}}_{\text{diverges as } k\to\infty}\mathcal{Q}\mathcal{N}_0. \tag{7.73}$$

In Part1 section 3, we see that the two eigenvalues of $\mathcal{M}$ can be found by

$$\det(\gamma\mathcal{I} - \mathcal{M}) = 0, \tag{7.74}$$

[1] We have reviewed **eigenvectors** and **eigenvalues** of matrices in Part I, Section 3.4. Eigenvectors can be symbolized as a natural reference frame where the matrix reduces to a simpler description. A matrix is a linear operation that takes a vector in and builds another one. In the frame of reference of eigenvectors, eigenvectors are only rescaled, that is, their magnitude changes but not their orientation. The eigenvalues for these eigenvectors provide the scaling factor for the linear transformation of the eigenvectors by the matrix.

Table 7.5 Central Euler nuclear decay calculations for different sets of time intervals dt and decay constants λ. The total time here maxtime × dt = 20 for all calculations. The red cell shows the calculations that lay outside of the calculated stability field.

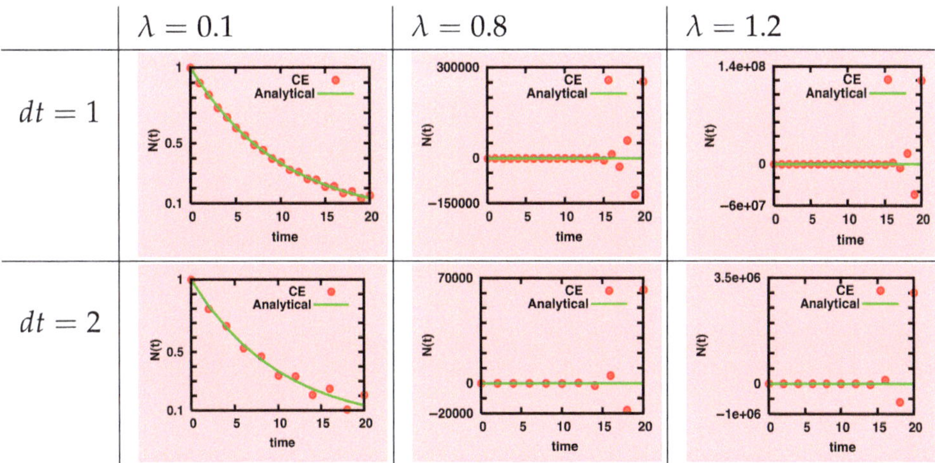

where det is the determinant and γ is either one of the two eigenvalues. We retrieve the quadratic equation

$$\gamma^2 + 2dt\lambda\gamma - 1 = 0. \tag{7.75}$$

The quadratic equations admits two roots, γ_1 and γ_2,

$$\gamma_{1,2} = \lambda dt\left(-1 \pm \sqrt{1 + \frac{1}{\lambda^2 dt^2}}\right). \tag{7.76}$$

Considering γ_2,

$$\gamma_2 < -\lambda dt - \sqrt{\lambda^2 dt^2 + 1} = -1 + \underbrace{f(\lambda dt)}_{<0} < -1. \tag{7.77}$$

So we find that $|\gamma_2| > 1$. For a large number of time steps $k \to \infty$, we find that the amplitude of $\mathcal{N}_k$ diverges and becomes unbounded irrespective of the choice of time step dt. In conclusion, we showed that the CE scheme for the nuclear decay equation is always unstable for any choice of time step dt. We refer to this sad case as *unconditional unstable.*

7.11 **Exercises**

1. Use the method described in Section 7.7 to derive a second-order scheme for d^2N/dx^2. Identify the leading term of the truncation error.
2. Write a Matlab script that takes in three inputs, the duration of a run `maxtime = maxiter*dt`, the time step `dt`, and the decay constant `lambda` and uses the FE approximation to solve the nuclear decay equation. Characterize the error $\epsilon = |N(t) - N_{FE}(t)|$, where

N is the analytical solution and N_{FE} is the approximation. Fix the decay constant to $\lambda = 0.1$, the duration of the run to 10, and use different dt values (0.01; 0.05; 0.1; 0.5; 1; 5; 10). Show the error ϵ (y-axis) against the choice of dt (x-axis) in a log–log plot. What is the value of the slope you obtain?

3. Repeat the same calculations for BE and plot it with the FE approximation. What do you observe?

Chapter 8

What controls convergence? What relates convergence and stability?

Chapter 7 introduced the Finite Difference Equation (FDE) approximation method and introduced three schemes to solve for the nuclear decay equation. We observed different behaviors and saw that resorting to higher-order schemes does not necessarily guarantee convergence. Ultimately, we want a mathematical procedure that allows us to test if a scheme is convergent and under what conditions it should be used. Convergence means that the approximated solution can be made to approach the analytical solution as closely as needed by reducing the time step (for time-dependent ODEs). Chapter 7 also discussed stability, which is a *different* property of the FDE solution that tells us whether the FDE solution remains bounded for any number of time steps with a timestep dt. Convergence:

$$|N(t) - \mathcal{N}(t)| = \epsilon(dt) \rightarrow 0 \qquad \text{as } dt \rightarrow 0\ (dx \rightarrow 0), \tag{8.1}$$

while stability

$$|\mathcal{N}(kdt)| \leq C \qquad \text{for any } k, \tag{8.2}$$

where $\mathcal{N}$ is the solution of the FDE, N is the analytical solution, and C is a finite positive constant. Stability is a necessary but insufficient condition for convergence, and this is easy to see because stability does not provide any information about the error between approximation and the analytical (true) solution.

8.1 Lax–Richtmyer Equivalence Theorem

In this section, we introduce a vulgarization of the Lax–Richtmyer theorem and its applicability for the convergence of numerical approximations to differential equations. The goal here is not to go over the details of the beautiful theory that was developed in the 1950s by both authors and the interested reader is referred to the original contribution [16] for more details. In 1956, Peter Lax and Robert Richtmyer showed that convergence is guaranteed if a finite difference scheme satisfies simultaneously two conditions:

1. Stability
2. Consistency.

Consistency has not been defined yet. Consistency is a property of the FDE, that tells us that if the discretization is refined (here $dt \rightarrow 0$), the FDE $\rightarrow$ Differential Equation (here ODE, but later PDE). In other words, it tells us that the truncation error is well behaved, that is, it goes to 0 if $dt \rightarrow 0$. In a very simplified way, we can consider consistency as a property that depends mostly on truncation errors, while stability deals generally with the accumulation of round-off errors over successive

Introduction to Numerical Modeling in Earth and Planetary Sciences, Christian Huber, Oxford University Press.
© Christian Huber (2025). DOI: 10.1093/oso/9780198802716.003.0009

time steps. Let us define again $N(t)$ as the analytical solution, $\mathcal{N}(t)$ as the solution of the FDE (contains both round-off and truncation errors), and finally $\mathcal{N}^*(t)$ as an ideal numerical solution that is free from round-off error. Under these conditions, *convergence* means that, as $dt \to 0$,

$$|N - \mathcal{N}| \to 0, \tag{8.3}$$

we define stability as

$$|\mathcal{N} - \mathcal{N}^*| < C_1|\Delta t|^a, \tag{8.4}$$

where $a > 0$ and $C_1 > 0$ are constants. Finally, consistency tells us about the behavior of truncation errors alone,

$$|N - \mathcal{N}^*| < C_2|dt|^b, \tag{8.5}$$

with $b, C_2 > 0$ as well. We can provide a very simplified idea of the proof of the theorem using the definition of convergence, where we can find that

$$|N - \mathcal{N}| = |\left(N - \mathcal{N}^*\right) + \left(\mathcal{N}^* - \mathcal{N}\right)|. \tag{8.6}$$

Using the triangle inequality, we get

$$|N - \mathcal{N}| \le |N - \mathcal{N}^*| + |\mathcal{N}^* - \mathcal{N}| < C_1|dt|^a + C_2|dt|^b. \tag{8.7}$$

As the time step $dt \to 0$, the right-hand side $\to 0$ because $a, b > 0$ and we have

$$|N - \mathcal{N}| \to 0 \qquad dt \to 0, \tag{8.8}$$

which "proves" that consistency and stability provide necessary and sufficient conditions for convergence.

8.2 Back to nuclear decay

The Lax–Richtmyer equivalence theorem offers us a simple diagnostic tool to understand the performance of our FDE approximations for the nuclear decay equation. Convergence, that is, that the approximated solution can be made as accurate as needed by decreasing dt, requires both stability and consistency. We discussed the stability of the Forward (FE), Backward (BE), and Centered Euler (CE) schemes in Chapter 7, and we need now to consider consistency. An FDE scheme is consistent if the truncation error $E^T \to 0$ as $dt \to 0$. The truncation error for each scheme is

$$E^T_{\text{FE}} = \frac{1}{2}\frac{d^2N}{dt^2}|_t dt + \ldots + \frac{1}{n!}\frac{d^nN}{dt^n}|_t dt^{n-1} + \ldots \tag{8.9}$$

$$E^T_{\text{BE}} = -dt\left[\frac{1}{2}\frac{d^2N}{dt^2}|_t + \lambda\frac{dN}{dt}|_t\right] + \ldots + \frac{1}{n!}\frac{d^nN}{dt^n}|_t dt^{n-1} + \ldots \tag{8.10}$$

$$E^T_{\text{CE}} = \frac{1}{3}\frac{d^3N}{dt^3}|_t dt^2 + \ldots + \frac{1}{n!}\left(1 - (-1)^n\right)\frac{d^nN}{dt^n}|_t dt^{n-1} + \ldots. \tag{8.11}$$

We find that as $dt \to 0$, all $E^T \to 0$, the three schemes are therefore consistent, because the FDE $\to$ ODE as $dt \to 0$. The stability analyses presented in the previous chapter reveal that FE and BE are respectively conditionally and unconditionally stable, while CE is unconditionally unstable. To summarize, we have

- FE: conditionally convergent. Converges if $-1 \leq 1 - \lambda dt \leq 1$.
- BE: unconditionally convergent.
- CE: unconditionally divergent (unstable for any choice of dt).

As a conclusion to the discussion of finite difference schemes for the nuclear decay equation, we stress that the conclusions drawn from each scheme here reflect the behavior of these approximations for the exponential decay law only. The CE scheme can outperform both FE and BE schemes when applied to a different ODE. Nevertheless, this example introduces some of the important features and concepts that remain relevant throughout our analysis of numerical schemes to solve ordinary and partial differential equations.

8.3 Exercises

1. Consider a new scheme for the nuclear decay

$$\frac{N((k+1)\Delta t) - N((k-1)\Delta t)}{2\Delta t} = -\frac{\lambda}{2}\left(N((k+1)\Delta t) + N(k\Delta t)\right). \tag{8.12}$$

   Use the Taylor series to get the order of the scheme and the first terms of the truncation error. Derive the stability condition for the scheme and discuss convergence.

Chapter 9

Box models: From single to multiple coupled ODEs

In this chapter, we continue our study of ordinary differential equations (ODEs) and introduce multiple coupled ODEs. We start with a set of definitions for the physical systems we want to solve, namely how these systems are coupled to their environment (open vs. closed systems) and we discuss the linearity of the set of ODEs. Coupled ODEs are often referred to as box models and are routinely used in Earth sciences to model the conservation of some quantities exchanged by different reservoirs. In general, these models admit time as their only differentiation variable. Box models are commonly applied to study elemental (or isotopic) cycles. Interested readers are referred to Francis Albarède's *Introduction to Geochemical Modeling* [1] for nice discussions and applications of box models in geochemistry.

9.1 Open versus closed systems

Similarly to thermodynamics, we define open and closed systems based on the ability of the reservoirs to exchange mass or energy with their environment (beyond the other reservoirs considered). For example, let us consider the Atlantic Ocean as one reservoir and track the amount of water in the ocean as the time-dependent variable of interest. Our goal is then to write down a set of ODEs that relates the evolution of the total mass of water in the Atlantic Ocean to other reservoirs also considered, for example, the Pacific Ocean, the Arctic Ocean, and the Mediterranean Sea. If all mass exchanges that need to be considered occur among the four reservoirs modeled, then we would refer to our water conservation model as a closed system, that is, the mass of water summed over these four reservoirs does not change over time. In reality, in this specific example, there are several other sources and sinks of mass that are not taken into account. For instance, the basins considered are also exchanging mass with other oceans and continental masses through river outputs. Oceans and atmosphere also exchange water through evaporation (sink) and precipitation (source). We can add parameterized sources and sinks to each of our four reservoirs to correct for these contributions, but the overall box model will become open, that is, the total mass of water summed over the four reservoirs can vary over time because of mass fluxes in and out of the system. Mathematically, if we define $A_i(t)$, $i = 1\ldots4$ to be the mass of water in the four reservoirs, a closed system is defined as $\sum_{i=1}^{4} A_i(t) = C$, where C is a constant. We represent graphically box models with charts where each reservoir (box) is connected to others through arrows (see Figure 9.1). As a general rule, if even a single box is connected to an arrow that does not relate it to another box in the model, the system is said to be open.

Introduction to Numerical Modeling in Earth and Planetary Sciences, Christian Huber, Oxford University Press.
© Christian Huber (2025). DOI: 10.1093/oso/9780198802716.003.0010

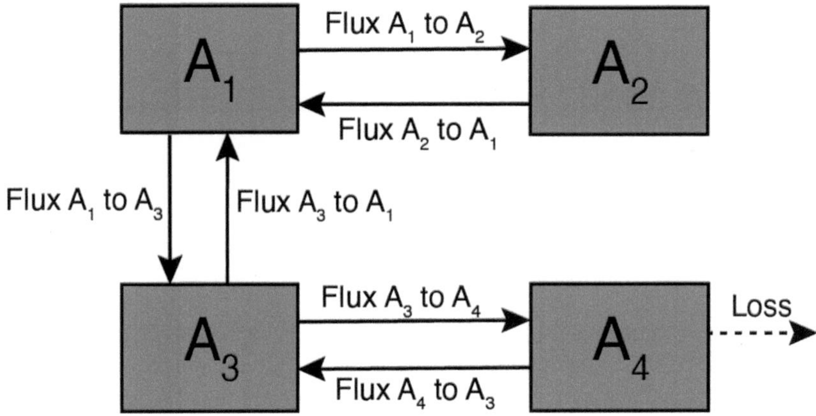

Figure 9.1 Schematic illustration of an open box model. Note that not all boxes are directly related to each other and that box 4 is losing mass to the environment (outside of system).

Using the most generic description of box models, we write a set of $k = 4$ ODEs for our reservoirs A_i so that

$$\frac{dA_1}{dt} = \text{Fluxes}_{\text{to } A_1} - \text{Fluxes}_{\text{out of } A_1} \tag{9.1}$$

$$\frac{dA_2}{dt} = \text{Fluxes}_{\text{to } A_2} - \text{Fluxes}_{\text{out of } A_2} \tag{9.2}$$

$$\frac{dA_3}{dt} = \text{Fluxes}_{\text{to } A_3} - \text{Fluxes}_{\text{out of } A_3} \tag{9.3}$$

$$\frac{dA_4}{dt} = \text{Fluxes}_{\text{to } A_4} - \text{Fluxes}_{\text{out of } A_4}. \tag{9.4}$$

We can read the fluxes coming into each reservoir as sources and the fluxes out of each reservoirs as sinks. Although this is a correct description, it introduces some cumbersome identification and we prefer to use the words "sources" and "sinks" only for mass exchanges with the environment (open systems). The main motivation behind this definition is that fluxes from one reservoir to another are simultaneously sinks and sources for the first and the latter, respectively, and we want to avoid confusion with notation as much as possible. By definition the units of the fluxes are [A/time].

Section 9.2 discusses how the fluxes can be expressed in terms of the conserved quantities of interest, A_is, to construct a set of closed-form expressions (ODEs) that can be solved analytically or numerically.

9.2 Linear box models

The simplest class of box models are linear, that is, the mass conservations are linear. In other words, if we rescale each of the reservoirs by, for example, a factor of 2 and do the same with the fluxes, then we retrieve the exact same evolution for each reservoir. The linearity of the system of differential equations that make up the box model also allows for the existence of analytical solutions.

9.2.1 Relationship between flux and conserved quantities

Fluxes out of a reservoir have to depend on the mass available in that reservoir. Let us take for example reservoir A_1 and assume that the total flux out of A_1 is constant ($= c_1$). Then

$$\frac{dA_1}{dt} = \text{Fluxes}_{\text{to } A_1} - c_1, \tag{9.5}$$

but this equation is problematic in that there is nothing to prevent the mass of the reservoir from becoming negative if $|c_1| > |\text{Fluxes}_{\text{to } A_1}|$ for a sufficiently long time (depending on the initial mass in A_1 as well). In order to avoid an unphysical behavior of that sort, we must make sure that the fluxes leaving each of the reservoirs decrease when the mass of these reservoirs decrease, that is, mass and fluxes should be positively correlated.

The most simple correlation is a linear relationship between the fluxes and the mass of the reservoir from which they originate. Taking two boxes A and B as an example and introducing the notation that F_{A-B} denotes the flux of mass from reservoir A to B, we can write in this case

$$F_{A-B} = k_{A-B}A(t), \tag{9.6}$$

where k_{A-B} is a rate constant (independent of $A(t)$ and with units of 1/time).

Using this notation, a generic three boxes (A, B, C) closed system are written as

$$\begin{aligned}
\frac{dA}{dt} &= -(k_{A-B} + k_{A-C})A(t) + k_{B-A}B(t) + k_{C-A}\,C(t) \\
\frac{dB}{dt} &= -(k_{B-A} + k_{B-C})\,B(t) + k_{A-B}A(t) + k_{C-B}\,C(t) \\
\frac{dC}{dt} &= -(k_{C-A} + k_{C-B})\,C(t) + k_{A-C}A(t) + k_{B-C}\,B(t),
\end{aligned} \tag{9.7}$$

where the constants k_{X-Y} are assumed positive. This linear system can be written in a matrix form defining the vector of reservoirs $\mathcal{V} = (A, B, C)^{T}$[1] and the matrix $\mathcal{M}$

$$\mathcal{M} = \begin{pmatrix} -(k_{A-B} + k_{A-C}) & k_{B-A} & k_{C-A} \\ k_{A-B} & -(k_{B-A} + k_{B-C}) & k_{C-B} \\ k_{A-C} & k_{B-C} & -(k_{C-A} + k_{C-B}) \end{pmatrix}, \tag{9.8}$$

the conservation (Eqs. 9.7) become

$$\frac{d\mathcal{V}}{dt} = \mathcal{M}\mathcal{V}. \tag{9.9}$$

The matrix $\mathcal{M}$ in (Eq. 9.8) has interesting properties that result from the structure of the conservation equations. First, for a closed system like this particular example, each column of $\mathcal{M}$ sums

[1] ^{T} refers to the transpose of the vector $\mathcal{V}$.

to 0 to enforce mass balance. Second, the diagonal is strictly negative while the off-diagonal components are positive (by construction $ks > 0$). Finally, $\mathcal{M}$ is symmetric only if the opposite rates match $k_{i-j} = k_{j-i}$, which does not necessarily means that the fluxes are balanced (which depends on the mass in the respective reservoirs). Some elements of the matrix may in certain conditions be null if two boxes are not exchanging mass directly, but overall we are left with a square matrix of rank N (N is the number of reservoirs) that is generally not sparse and not symmetric.

For an open system where mass is exchanged between reservoirs and the environment that surrounds the system, the governing equations can be modified slightly to

$$\frac{d\mathcal{V}}{dt} = \mathcal{M}\mathcal{V} + \mathcal{E}, \tag{9.10}$$

where

$$\mathcal{E} = \begin{pmatrix} S_A - L_A \\ S_B - L_B \\ S_C - L_C \end{pmatrix}, \tag{9.11}$$

and S_i and L_i refer respectively to the external sources and sinks for the reservoir i.

9.2.2 **Analytical solutions**

Under the assumption that the fluxes in and out of reservoirs are linearly related to the mass of the reservoirs, it is possible to derive an analytical solution for the mass conservation of all reservoirs simultaneously relying on basic linear algebra. We start with a closed box model with N boxes and a square rate matrix $\mathcal{M}$ of size $N \times N$ similar to the definition in (Eq. 9.8). Using the same notation as in Section 9.2.1, we can decompose the vector of reservoir masses $\mathcal{V}$ on any basis of linearly independent vectors (the basis also contains N independent vectors $\mathbf{v}_j, j = 1...N$)

$$\mathcal{V}(t) = \sum_{j=1}^{N} a_j(t)\mathbf{v}_j, \tag{9.12}$$

where the a_js are coefficients that we need to determine. An obvious choice for the vector basis $\{\mathbf{v}_1, \mathbf{v}_2, ..., \mathbf{v}_N\}$ is the set of N linearly independent eigenvectors of the matrix $\mathcal{M}$ and we use this specific basis to write our solution vector $\mathcal{V}$. Introducing (Eq. 9.12 into Eq. 9.8), we obtain

$$\sum_{j=1}^{N} \frac{da_j}{dt}\mathbf{v}_j = \sum_{j=1}^{N} a_j\mathcal{M}\mathbf{v}_j = \sum_{j=1}^{N} a_j\lambda_j\mathbf{v}_j, \tag{9.13}$$

where $\lambda_j, j = 1...N$ are the eigenvalues of $\mathcal{M}$ for the eigenvectors $\mathbf{v}_j$. The eigenvectors for real-valued matrices are linearly independent, hence comparing the last two terms of the previous equations we obtain a set of linear ODEs for the coefficients a_j

$$\frac{da_j}{dt} = \lambda_j a_j, \quad j = 1...N. \tag{9.14}$$

These equations are similar to the nuclear decay equation with decay rates $-\lambda_j$ and we get

$$a_j(t) = a_j(0)\exp(\lambda_j t). \tag{9.15}$$

Substituting these coefficients back into (Eq. 9.12) yields

$$\mathcal{V}(t) = \sum_{j=1}^{N} a_j(0)\exp(\lambda_j t)\mathbf{v}_j. \tag{9.16}$$

A unique solution exists for the coefficients $a_j(0)$ if an initial condition for $\mathcal{V}$ is provided, here $\mathcal{V}_0$. This can be seen by setting the time to $t = 0$ in this equation

$$\mathcal{V}_0 = \sum_{j=1}^{N} a_j(0)\mathbf{v}_j. \tag{9.17}$$

Grouping the coefficients $a_j(0)$ into a vector $\mathcal{A}_0$, we can invert (Eq. 9.16) to get

$$\mathcal{A}_0 = \mathcal{K}^{-1}\mathcal{V}_0, \tag{9.18}$$

where the columns of the square $(N \times N)$ matrix $\mathcal{K}$ consist of the eigenvectors $\mathbf{v}_1, \ldots, \mathbf{v}_N$. The analytical solution to the linear closed-box model is finally

$$\mathcal{V}(t) = \sum_{j=1}^{N} \left(\mathcal{K}^{-1}\mathcal{V}_0\right)_j \exp(\lambda_j t)\mathbf{v}_j. \tag{9.19}$$

The same general approach can be followed for open linear box models. However, there are two necessary modifications. First, the sink or loss terms to the environment L_j that are also linearly related to the mass in the *jth* reservoir are added to the diagonal of matrix $\mathcal{M}$ and the source terms S_j, which are independent of the mass in each reservoir, provide nonhomogeneous terms to the set of coupled linear differential equations. Using the standard procedure for nonhomogeneous linear ODEs, we must seek a particular solution that will be combined with the homogeneous solution provided by (Eq. 9.19). Defining a vector of source terms $\mathcal{S} = (S_1, \ldots, S_N)^T$, if the source terms do not depend on time, a suitable particular solution would be

$$\mathcal{V}_p = -\mathcal{M}^{-1}\mathcal{S}. \tag{9.20}$$

9.2.3 Example 1: Nuclear decay chain—uranium series

We started our discussion on the application of the finite difference method with the simple nuclear decay equation. We pursue this example later, but now focus our attention to the nuclear decay series of certain radiogenic isotopes. A simple example of such a series is the uranium decay series for ^{238}U

$$^{238}\mathrm{U} \rightarrow {}^{234}\mathrm{Th} \rightarrow {}^{234}\mathrm{Pa} \rightarrow {}^{234}\mathrm{U} \rightarrow {}^{230}\mathrm{Th} \rightarrow {}^{226}\mathrm{Ra} \rightarrow {}^{222}\mathrm{Rn} \tag{9.21}$$

$$\rightarrow {}^{218}\mathrm{Po} \rightarrow {}^{214}\mathrm{Pb} \rightarrow {}^{214}\mathrm{Bi} \rightarrow {}^{214}\mathrm{Po} \rightarrow {}^{210}\mathrm{Pb} \rightarrow {}^{210}\mathrm{Bi} \rightarrow {}^{210}\mathrm{Po} \rightarrow {}^{206}\mathrm{Pb}. \tag{9.22}$$

The two main decay types here are alpha decays, which are accompanied with a loss of 4 a.m.u., and beta decays, which represent a conversion from proton to neutron (or vice versa) and do not modify the mass of the nuclide. Each step of the reaction series proceeds at a different rate, that is, different half-lives that are listed in Table 9.1.

In the table the diagonal components show the half-life of that element and the decay leads to the formation of the daughter element located one row down (or one column to the left). The half-lives span about 20 orders of magnitude, which will be problematic for modeling. In the context of a simple unique nuclear decay, we know that the accuracy and stability (for the Forward Euler, FE, method) depend on the product λ and the choice of time step dt. By definition, the decay constants can be derived from the half-lives in the following way

$$\frac{N_0}{2} = N_0 \exp(-\lambda t_{1/2}), \tag{9.23}$$

where N_0 is the initial concentration of radionuclides and $t_{1/2}$ is the half-life for the decay. From this equation we retrieve the relationship between the decay constant and the half-life of a radionuclide

$$\lambda = \frac{\ln(2)}{t_{1/2}}. \tag{9.24}$$

It is worth emphasizing here that the set of λ_j defined by (Eq. 9.24) are not the eigenvalues of the linear box models but here represent the half-lives of the different decay in the series. For convenience, in this section, we note the eigenvalues of the linear system of ODEs with the letter γ. An illustration of the first few steps of the box model for the decay series is shown in Figure 9.2, where the fluxes from one box to the next $F_{\text{A-B}}$ describe the decay of element A into B at a rate

$$F_{\text{A-B}} = \lambda_{\text{A-B}} A(t), \tag{9.25}$$

where $A(t)$ is the time-dependent concentration of the parent nuclide and

$$\lambda_{\text{A-B}} = \frac{\ln(2)}{t_{1/2}^{\text{A-B}}}. \tag{9.26}$$

Recasting the decay constant into a square matrix of the following form

$$\mathcal{M} = \begin{pmatrix} -\lambda_{U-Th} & 0 & \dots & \\ \lambda_{U-Th} & -\lambda_{Th-Pa} & 0 & \dots \\ 0 & \lambda_{Th-Pa} & -\lambda_{Pa-U} & \dots \\ \dots & \dots & \dots & \dots \end{pmatrix}, \tag{9.27}$$

we can compute the analytical solution derived above (Eq. 9.19) to solve this linear system of differential equations and find that the concentration vector of nuclide concentrations $\mathcal{V} \equiv \left({}^{238}\text{U}, {}^{234}\text{Th}, \dots\right)$

$$\mathcal{V}(t) = \sum_{j=1}^{j=15} \left(\mathcal{K}^{-1}\mathcal{V}_0\right)_j \exp(\gamma_j t)\mathbf{v}_j, \tag{9.28}$$

where the γs are the eigenvalues of the matrix $\mathcal{M}$ associated with the eigenvectors $\mathbf{v}_j$.

Table 9.1 Half-lives for the radiogenic nuclides in the ^{238}U decay series, *y* refers to years, *m* to months, *d* to days, and *s* to seconds

	^{238}U	^{234}Th	^{234}Pa	^{234}U	^{230}Th	^{226}Ra	^{222}Rn	^{218}Po	^{214}Pb	^{214}Bi	^{214}Po	^{210}Pb	^{210}Bi	^{210}Po	^{206}Pb
^{238}U	4.5×10^9 y														
^{234}Th		24 d													
^{234}Pa			1.2 m												
^{234}U				248 ky											
^{230}Th					75 ky										
^{226}Ra						1622 y									
^{222}Rn							38 d								
^{218}Po								3.1 m							
^{214}Pb									27 m						
^{214}Bi										20 m					
^{214}Po											2×10^{-4} s				
^{210}Pb												22 y			
^{210}Bi													50 d		
^{210}Po														138 d	
^{206}Pb															

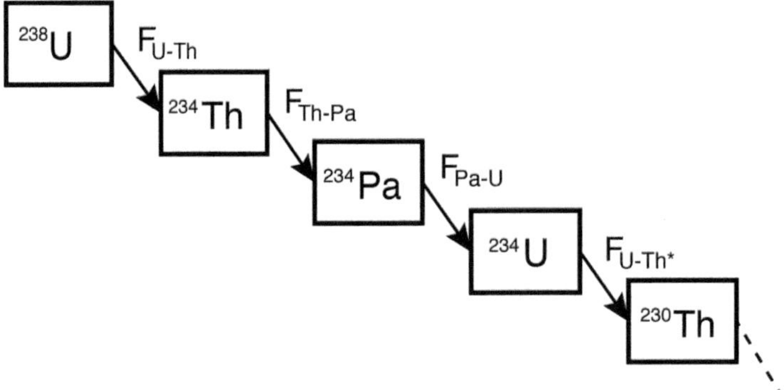

Figure 9.2 Illustration of the first part of the linear box model to represent one of uranium decay series.

Finite difference models for the nuclear decay series: Forward Euler (FE) method

Although analytical solutions exist for this particular case, we use it to illustrate the fundamental conditions for the numerical stability of linear (closed system) box models. The starting equations are

$$\begin{aligned}\frac{d^{238}\mathrm{U}(t)}{dt} &= -\lambda_{U-Th}\,{}^{238}\mathrm{U}(t)\\ \frac{d^{234}\mathrm{Th}(t)}{dt} &= +\lambda_{U-Th}\,{}^{238}\mathrm{U}(t) - \lambda_{Th-Pa}\,{}^{234}\mathrm{Th}(t)\\ \frac{d^{234}\mathrm{Pa}(t)}{dt} &= +\lambda_{Th-Pa}\,{}^{234}\mathrm{Th}(t) - \lambda_{Pa-U}\,{}^{234}\mathrm{Pa}(t)\\ \ldots &= \ldots\end{aligned} \tag{9.29}$$

Using the FE discretization for the time derivatives leads to the marching equations

$$\begin{aligned}{}^{238}\mathrm{U}(t+dt) &= (1-\lambda_{U-Th}dt)\,{}^{238}\mathrm{U}(t) \qquad (9.30)\\ {}^{234}\mathrm{Th}(t+dt) &= \lambda_{U-Th}dt\,{}^{238}\mathrm{U}(t) + (1-\lambda_{Th-Pa}dt)\,{}^{234}\mathrm{Th}(t)\\ {}^{234}\mathrm{Pa}(t+dt) &= +\lambda_{Th-Pa}dt\,{}^{234}\mathrm{Th}(t) + (1-\lambda_{Pa-U}dt)\,{}^{234}\mathrm{Pa}(t)\\ \ldots &= \ldots\,.\end{aligned}$$

In a matrix-vector format these equations reduce to

$$\mathcal{V}(t+dt) = \mathcal{M}_{FE}\mathcal{V}(t), \tag{9.31}$$

where $\mathcal{V}$ is again the vector of nuclide concentration and $\mathcal{M}_{FE}$ is

$$\mathcal{M}_{FE} = \begin{pmatrix} 1-\lambda_{U-Th}dt & 0 & \ldots & \\ \lambda_{U-Th}dt & 1-\lambda_{Th-Pa}dt & 0 & \ldots \\ 0 & \lambda_{Th-Pa}dt & 1-\lambda_{Pa-U}dt & \ldots \\ \ldots & \ldots & \ldots & \ldots \end{pmatrix}. \tag{9.32}$$

The pseudo-code to solve for the FE finite difference approximation for the nuclear decay series example looks like

```
V0=(U_238_ini,Th_234_ini, ...); % Initial condition
V(1)=V0;                    % Applies Initial cond.
dt=...;          % Choice of time step
N=15;                 % Number of nuclides considered
M_FE=zeros(N,N);
M_FE(1,1)=1-lambda_U_Th*dt;
M_FE(2,1)=lambda_U_Th*dt;
...
for time=1:maxtime-1                  % Marching equation
    V(time+1)=M_FE*V(time);   % FE FDE
end
```

It is now time to consider the stability of this scheme. We saw that the FE model for the single nuclear decay had a conditional stability that depended on the choice of time step dt and the decay constant λ. The same is true here, although we have more than one decay constant to consider. The first task is to write the vector of initial nuclide concentrations $\mathcal{V}(0)$ in terms of a linear combination of the eigenvectors $\mathbf{v}_k$ of the matrix $\mathcal{M}_{FE}$

$$\mathcal{V}(0) = \sum_k a_k \mathbf{v}_k. \tag{9.33}$$

Applying (Eq. 9.31), we get the marching system

$$\mathcal{V}(dt) = \mathcal{M}_{FE}\mathcal{V}(0) \tag{9.34}$$

$$\mathcal{V}(2dt) = \mathcal{M}_{FE}\mathcal{V}(dt) = \mathcal{M}_{FE}^2\mathcal{V}(0) \tag{9.35}$$

$$\dots = \dots \tag{9.36}$$

$$\mathcal{V}(ndt) = \mathcal{M}_{FE}^n\mathcal{V}(0). \tag{9.37}$$

Going back to the eigenvector decomposition of $\mathcal{V}(0)$ (Eq. 9.33), we find that

$$\mathcal{M}_{FE}^2\mathcal{V}(0) = \mathcal{M}_{FE}\left(\sum_k a_k\gamma_k\mathbf{v}_k\right) = \sum_k a_k\gamma_k^2\mathbf{v}_k, \tag{9.38}$$

where γ_k are the eigenvalues of $\mathcal{M}_{FE}$. Similarly,

$$\mathcal{M}_{FE}^n\mathcal{V}(0) = \sum_k a_k\gamma_k^n\mathbf{v}_k. \tag{9.39}$$

From this simple development we see that the magnitude of the vector of concentration $\mathcal{V}$ is rescaled over time by factors that depend on powers of the eigenvalues γ_k of the matrix $\mathcal{M}_{FE}$.

Stability again refers to the property that the numerical approximation of the solution of the system of differential equations remains bounded for all time. In other words, as long as none of the factors $a_k\gamma_k^n$ diverge for any choice of n, the solution remains bounded and the scheme is deemed stable. Because $\mathcal{M}_{FE}$ depends on the set of decay constants and dt, we can readily see that stability is controlled by both as expected. Bounded solutions are guaranteed if each of the eigenvalues is such that $|\gamma_k| \leq 1$. The problems of numerical stability for the FE scheme arise because of the radioactive elements with a short half-life (large decay constant), that is, ^{214}Po. To balance this rapid decay, we must use a short time step, the value of which is determined in the following exercise. We also see in the exercise that the size of the time step required to satisfy stability is such that it makes any useful calculation of the decay series evolution prohibitively heavy to solve numerically. It is therefore one of the cases where a different finite difference scheme can make a huge difference.

Using the example of the simple nuclear decay of a single nuclide, we expect that a Backward Euler (BE) discretization of the time derivative would offer unconditional stability (i.e., does not depend on the choice of time step dt). Applying the BE method to our set of ODEs, we retrieve

$$\begin{aligned} {}^{238}\mathrm{U}(t+dt)\,(1+\lambda_{U\text{-}Th}dt) &= {}^{238}\mathrm{U}(t) \\ {}^{234}\mathrm{Th}(t+dt)(1+\lambda_{Th\text{-}Pa}dt) - {}^{238}\mathrm{U}(t+dt)\lambda_{U\text{-}Th}dt &= {}^{234}\mathrm{Th}(t) \\ {}^{234}\mathrm{Pa}(t+dt)(1+\lambda_{Pa\text{-}U}dt) - {}^{234}\mathrm{Th}(t+dt)\lambda_{Th\text{-}Pa}dt &= {}^{234}\mathrm{Pa}(t) \\ \ldots &= \ldots \end{aligned} \tag{9.40}$$

This equation can be written in a compact matrix-vector form

$$\mathcal{M}_{BE}\mathcal{V}(t+dt) = \mathcal{V}(t), \tag{9.41}$$

where the evolution matrix $\mathcal{M}_{BE}$ is defined as

$$\mathcal{M}_{BE} = \begin{pmatrix} 1+\lambda_{U\text{-}Th}dt & 0 & \ldots & \\ -\lambda_{U\text{-}Th}dt & 1+\lambda_{Th\text{-}Pa}dt & 0 & \ldots \\ 0 & -\lambda_{Th\text{-}Pa}dt & 1+\lambda_{Pa\text{-}U}dt & \ldots \\ \ldots & \ldots & \ldots & \ldots \end{pmatrix}. \tag{9.42}$$

Note the subtle differences between the matrices $\mathcal{M}_{BE}$ and $\mathcal{M}_{FE}$. Additionally, $\mathcal{M}_{FE}$ is the matrix that allows to solve for a system of marching equations while marching equations for the BE method require taking the inverse of $\mathcal{M}_{BE}$

$$\mathcal{V}(t+dt) = (\mathcal{M}_{BE})^{-1}\,\mathcal{V}(t). \tag{9.43}$$

The similarities with the marching equation for the FE allow us to recognize that the magnitude of the eigenvalues of the matrix $\mathcal{K} \equiv (\mathcal{M}_{BE})^{-1}$ control the stability of the scheme. We need that every eigenvalue γ_k of $\mathcal{K}$ is such that $|\gamma_k| \leq 1$. The stability condition is then mostly identical to that of the FE, although note that $\gamma_k \propto 1/(1+\lambda_k dt)$ instead of $\gamma_k \propto (1-\lambda_k dt)$ for the FE discretization. This subtle difference is a huge asset because it provides unconditional stability to the scheme, which we test in the following exercises.

9.2.4 **Exercises**

1. Use Table 9.1 to build the matrices $\mathcal{M}_{FE}$ and $\mathcal{M}_{BE}$. Use the function **eig** in Matlab to compute the eigenvalues of each matrices. In the light of the discussion on stability just above, what do you expect to see when applying the FE or BE discretization? What is the maximum duration of a time step that will ensure stability?
2. Compare the solutions to your numerical model with time (using the BE method) with the analytical solution.

9.2.5 **Example 2: Ca isotopes in the oceans**

The model we implement for the evolution of Ca isotopes in oceans is based on De La Rocha and DePaolo [9]. We refer the reader to the original study for more detailed discussion of the model and assumptions. The goal of such approaches is to use the Ca isotopic composition of sediments to fingerprint changes in ocean composition driven by climate-driven forcings. De La Rocha and DePaolo use a box model approach to Ca isotopes to build time series of past continental weathering rates to reconstruct atmospheric composition during the Eocene.

In the model they propose, the ocean Ca composition is driven by two general sources and one sink. The two sources are the weathering of mid ocean ridge basalts and continental weathering and the sink is the precipitation of carbonates. The archive we can interrogate to assess the relative fluctuations in continental weathering flux over time (which is strongly controlled by sea level change and therefore changes in volume of the icesheets) is the sediments deposited at the bottom of the ocean. Because Ca is a stoichiometric component of the carbonates in the sediments, we have to resort to its stable isotopes to get back at the relative importance of weathering versus precipitation.

The idea here is that each of flux (weathering, precipitation) has a specific Ca isotopic composition. The two isotopes of Ca that we focus on are ^{40}Ca and ^{44}Ca, which respectively account for about 97% and 1% of Ca at the Earth's surface. Generally, the isotopic composition of a sample is defined by the ratio of the rare to the more abundant isotope, here

$$^{44}R = \frac{^{44}Ca}{^{40}Ca}, \tag{9.44}$$

this number is small in general and it is convenient to introduce the delta notation, which compares the sample to a reference

$$\delta^{44}Ca = \left[\left(\frac{^{44}R_{\text{sample}}}{^{44}R_{\text{ref}}}\right) - 1\right] \times 1000, \tag{9.45}$$

where $^{44}R_{\text{x}}$ is the isotopic ratio of x. We simplify the notation and use

$$\delta^{44}x = \left[\left(\frac{^{44}R_{\text{x}}}{^{44}R_{\text{ref}}}\right) - 1\right] \times 1000, \tag{9.46}$$

with x representing the provenance of the sample and $\delta^{44}x$ its Ca isotopic composition in delta notation.

The precipitation of carbonate (flux of Ca out of the ocean) by micro-organism, in the ocean is assumed to be lighter (enriched in ^{40}Ca) than the ocean Ca composition by a constant shift Δ_{sed}

$$\Delta_{\mathrm{sed}} = \delta sed - \delta sw, \tag{9.47}$$

where δsed and δsw refer respectively to the isotopic composition of the precipitated sediments and that of the seawater from which sediments precipitated. The mass of Ca in the oceans N_{Ca} is (currently) about 1.4×10^{19} grams. The amount of Ca dissolved in the oceans changes in our model according to the balance between fluxes

$$\frac{dN_{Ca}}{dt} = F_{wc} + F_{MOR} - F_{sed}, \tag{9.48}$$

with the subscripts *wc* and *MOR* referring to the fluxes from the weathering of continents and mid ocean ridge basalts. Now the Ca isotopic composition of the oceans varies according to the fluxes and their associated Ca isotopic composition

$$\frac{dN_{Ca}\delta sw}{dt} = F_{wc}\delta wc + F_{MOR}\delta MOR - F_{sed}\delta sed. \tag{9.49}$$

Using the chain rule, we get an expression for the evolution of the seawater composition

$$N_{Ca}\frac{d\delta sw}{dt} = F_{wc}\delta wc + F_{MOR}\delta MOR - F_{sed}\delta sed - \delta sw\frac{dN_{Ca}}{dt}, \tag{9.50}$$

and replacing the last derivative with the mass balance expression for total Ca in the ocean above, we get

$$N_{Ca}\frac{d\delta sw}{dt} = F_{wc}(\delta wc - \delta sw) + F_{MOR}(\delta MOR - \delta sw) - F_{sed}(\delta sed - \delta sw). \tag{9.51}$$

For simplicity, following De La Rocha and DePaolo, we introduce a general weathering flux which includes both weathering fluxes

$$F_w(\delta w - \delta sw) = F_{wc}(\delta wc - \delta sw) + F_{MOR}(\delta MOR - \delta sw), \tag{9.52}$$

which finally leads to the equation published by De La Rocha and DePaolo after substituting for Δ_{sed}

$$N_{Ca}\frac{d\delta sw}{dt} = F_w(\delta w - \delta sw) - \delta sw) - F_{sed}\Delta_{\mathrm{sed}}, \tag{9.53}$$

with the total dissolved Ca in the ocean following

$$\frac{dN_{Ca}}{dt} = F_w - F_{sed}. \tag{9.54}$$

9.2.6 Exercises

We assume a fixed (in time) isotopic composition for the weathering fluxes. Lastly, we assume that the precipitation flux $F_{sed} = 2.9 \times 10^{12}$ g/yr remains fixed over the last 80 million years (strong assumption). The composition of the weathering flux is fixed at $\delta w = -0.44$, and the shift in isotopic composition between the sediments and seawater is $\Delta_{\text{sed}} = -1.3$. In this particular exercise, we do not reproduce the work of De La Rocha and DePaolo, but rather use their results, a linearized (by segments) relative weathering flux function and compare the modeled sediment Ca isotopic composition with the real data they analyzed.

1. Use the linearized relative weathering flux

$$\frac{F_w}{F_{sed}} = \begin{cases} 2.59 \times 10^{-8} + 3.041 & -80\text{Ma} < t \le -76\text{Ma}, \\ -9.39 \times 10^{-9} + 0.357 & -76\text{Ma} < t \le -40\text{Ma}, \\ 7.25 \times 10^{-8} + 3.623 & -40\text{Ma} < t \le -34\text{Ma}, \\ -5.67 \times 10^{-10} + 1.137 & -34\text{Ma} < t \le -26\text{Ma}, \\ -7.86 \times 10^{-8} - 0.904 & -26\text{Ma} < t \le -22\text{Ma}, \\ 8.063 \times 10^{-8} + 2.65 & -22\text{Ma} < t \le -17.7\text{Ma}, \\ -6.22 \times 10^{-8} + 0.111 & -17.7\text{Ma} < t \le -14\text{Ma}, \\ -10 \times 10^{-9} + 0.87 & -14\text{Ma} < t \le 0\text{Ma}. \end{cases}$$

2. The sediment composition data ($\delta sed(t)$) is (from De La Rocha and DePaolo)

Age [Ma]	Composition (δsed)
−0.0194	−0.359
−3.99	−0.476
−11.5	−0.527
−15.6	−0.555
−17.7	−0.81
−19.9	−0.369
−20.5	−0.186
−25.9	−0.74
−31.76	−0.59
−34	−0.718
−36	−0.105
−38.25	−0.024
−39.76	−0.126
−44.8	−0.289
−64.9	−0.42
−76.2	−0.61

3. Solve the two coupled ODEs (Eqs. 9.53 and 9.54) with an FE scheme with a time step of 80 years. The initial time is −80 Ma and the simulation ends with the present (10^6 iterations).
4. Looking at your results and the data for the weathering flux, can you identify two possible major ice sheet growth periods?

5. Repeat the same calculations with a different fixed sedimentation rate of $F_{sed} = 2 \times 10^{12}$ gram/year. Can this be a reasonable flux, why?

9.2.7 Example 3: An idealized phosphorus cycle

This exercise is derived from the publications of Lasaga [15] and Albarède [1]. The model is linear and admits a general analytical solution (see Eq. 9.19). Nevertheless, we use it as an example for an application of the BE scheme to box models. Following the model described in Albarède [1], we assume six reservoirs (boxes) for phosphorus contain a steady mass of phosphorus listed in Table 9.2.

Figure 9.3 shows a diagram with the fluxes that link the different reservoirs. Under steady-state conditions, the finite fluxes between reservoirs are listed in Table 9.3 in units of 10^9 kg/year

9.2.8 Exercises

1. Use the fluxes listed in Table 9.3 to show that the system is at steady-state.
2. Derive rate constants from fluxes and steady-state mass in reservoirs.

Table 9.2 List of initial mass of phosphorus in each of the 6 boxes in the model, assuming the initial condition is a steady state.

Reservoir	**Mass [10^9 kg]**
Sediments (**1**)	4×10^9
Land (**2**)	2×10^5
Terrestrial biota (**3**)	3000
Oceanic biota (**4**)	138
Surface ocean (**5**)	2710
Deep ocean (**6**)	8.71×10^4

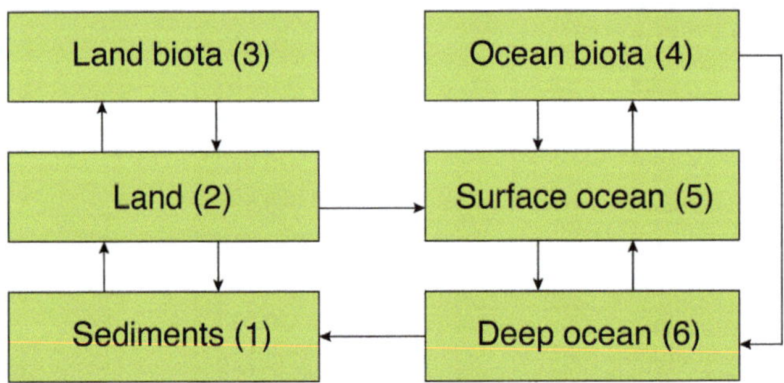

Figure 9.3 Schematic illustration of the phosphorus box model.

Table 9.3 Computed fluxes assuming that the masses in Table 9.2 are from steady state conditions.

$F_{1\text{-}2} = 20$	$F_{3\text{-}2} = 63.5$	$F_{5\text{-}6} = 17.7$
$F_{2\text{-}1} = 18.3$	$F_{4\text{-}5} = 998$	$F_{6\text{-}1} = 1.7$
$F_{2\text{-}3} = 63.5$	$F_{4\text{-}6} = 42$	$F_{6\text{-}5} = 58$
$F_{2\text{-}5} = 1.7$	$F_{5\text{-}4} = 1040$	

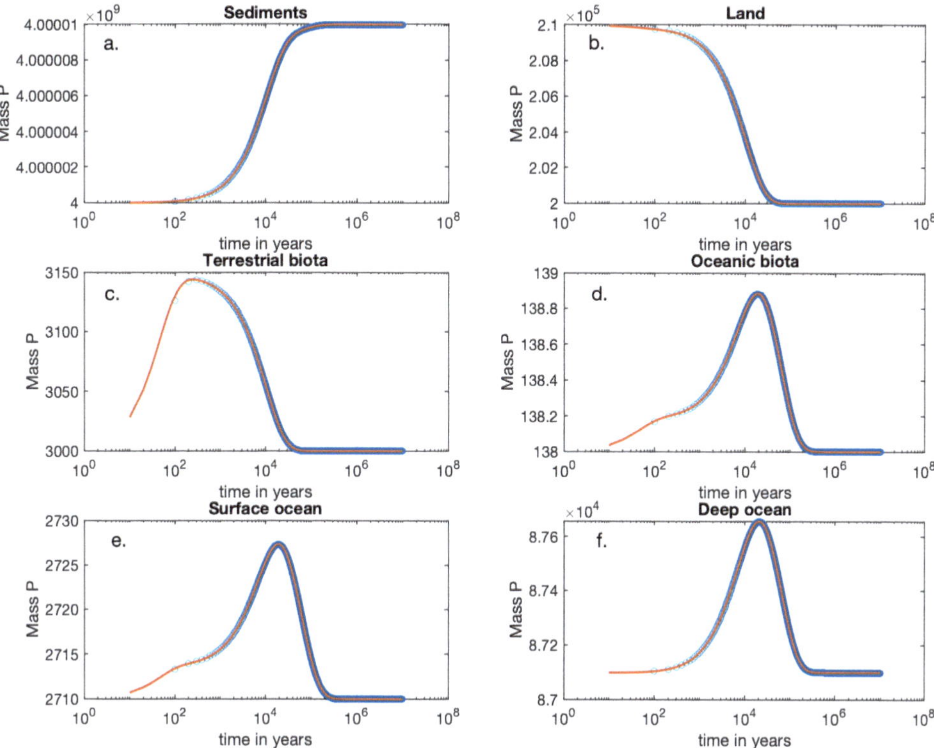

Figure 9.4 Backward Euler simulation of the phosphorus box model with an initial perturbation caused by the dumping of 10^{13} kg of phosphorus from fertilizers at time t = 0. The time step used is Δt = 10 years. The solid line shows the analytical solution computed with Eq. 9.19.

3. Use a BE method to solve these coupled ODEs, set a perturbation to the initial condition so that an additional 10^{13} kg of phosphorus is dumped on land (box 2) at the beginning of the run.
4. Compare your numerical solution to the analytical solution derived in (Eq. 9.19); see Figure 9.4 for the desired output.

9.3 Non-linear box models

9.3.1 Example 5: A simplified shallow magma chamber

Magma chambers host a wide variety of dynamical processes and are a good example of open systems (fluxes in and out). We use a simple box model approach to study the evolution of magma chambers according to the model proposed by Huppert and Woods [14]. Figure 9.5 provides a schematic illustration of the systems and of the fluxes in and out of the chamber.

We define ρ to be the average density (mass/volume) in the chamber, V as the volume of magma in the chamber, Q_i is the mass per unit time of magma added to the chamber from depth, and Q_o is the mass per unit time removed from the chamber by an eruption. The mass conservation of magma in the system then follows

$$\frac{d(\rho V)}{dt} = Q_i - Q_o. \tag{9.55}$$

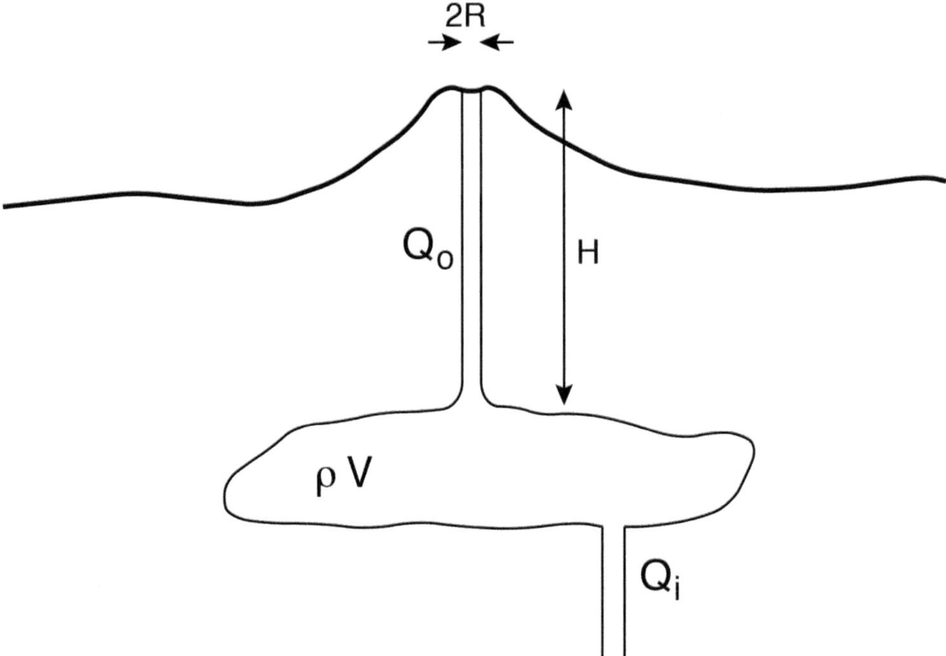

Figure 9.5 Schematic illustration of the magma chamber box model.

Following the assumptions of Huppert and Woods, we first assume that $Q_i = 0$, because the magma recharge rate is generally small compared to the output during an eruption. The left-hand side can be expanded into

$$\frac{d(\rho V)}{dt} = V\frac{d\rho}{dt} + \rho\frac{dV}{dt}. \tag{9.56}$$

For simplicity, the density of the magma is then assumed to be a sole function of pressure p and temperature T

$$\frac{d\rho}{dt} = \frac{dp}{dt}\frac{\partial\rho}{\partial p} + \frac{dT}{dt}\frac{\partial\rho}{\partial T}. \tag{9.57}$$

The last term is neglected because the thermal expansion coefficient of the magma is assumed to be small compared to its compressibility. Considering that the change of volume of the reservoir with time is elastic and depends solely on the pressure in the magma body, we get

$$\frac{1}{V}\frac{dV}{dt} = \frac{1}{\beta_r}\frac{dp}{dt}, \tag{9.58}$$

where β_r is the bulk modulus of the host rocks around the reservoir. With these constitutive models, we find that the mass conservation equation leads to

$$\left(\frac{1}{\rho}\frac{\partial\rho}{\partial p} + \frac{1}{\beta_r}\right)\frac{dp}{dt} = -\frac{Q_o}{\rho V}. \tag{9.59}$$

We can close the model by defining the equation of state of the magma ($\rho(p)$) and deriving a simple relationship to relate Q_o to the pressure-driven flux of magma through the conduit of length H.

Assuming a conduit with an effective radius R, the mass of magma leaving the chamber per unit time is

$$Q_o = \frac{\rho S(\pi R^2)^2}{H\mu}\left(p(t) - p_{\text{litho}}\right). \tag{9.60}$$

In this equation, $p_{\text{litho}} = \rho_r g H$ is the lithostatic pressure at the level of the chamber, ρ_r the average host rock density, μ the dynamic viscosity of the magma and S is a shape factor to account for the fact that the conduit is not necessarily perfectly smooth and cylindrical. The magma can host three different phases: the melt with density ρ_m; crystals with density ρ_c and mass fraction x; and gas bubbles with a density that is given by the ideal gas law $\rho_g = p(t)/R_g T$, where R_g is the gas constant. The mass fraction of gas bubbles is n and depends on the total amount of water in the magma (N = water dissolved + exsolved) and the solubility of water in the melt (Henry's law)

$$n(t) = N - s_s\sqrt{p(t)}(1 - x), \tag{9.61}$$

where s_s is Sievert's constant. For simplicity, we assume that N, s_s, and x remain constant during each calculation. The density of the three-phase magma is calculated according to

$$\rho(t) = \left[\frac{n(t)}{\rho_g(t)} + (1 - n(t))\left(\frac{x}{\rho_c} + \frac{1 - x}{\rho_m}\right)\right]^{-1}. \tag{9.62}$$

We can differentiate this equation with respect to pressure to retrieve the term $\partial\rho/\partial p$ in (Eq. 9.59)

$$\frac{\partial\rho}{\partial p} = \frac{n(t) + (1 - n(t))\alpha\frac{p}{R_g T} - \frac{(N - n(t))}{2}\left(1 - \frac{\alpha p}{R_g T}\right) + (1 - n(t))\frac{\alpha}{R_g T}}{R_g T\left(n(t) + (1 - n(t)\alpha\frac{p}{R_g T}\right)^2}, \tag{9.63}$$

with

$$\alpha = \frac{x}{\rho_c} + \frac{1 - x}{\rho_m}. \tag{9.64}$$

We see clearly from (Eqs. 9.59, 9.60, 9.61, 9.62 and 9.63) that the system is not linear with respect to density and pressure. It is therefore not possible to write this system of equations with matrix algebra and resort to an implicit (BE) scheme to solve it numerically. We first solve this system of equations with an FE time integration and leave for interested readers the implementation of a higher—order method such as Runge–Kutta integrations (explained later in this section). The problem setup for an FE discretization is straight-forward. The only two differential equations we have to discretize are (Eq. 9.59) and (Eq. 9.58), which leads to

$$p(t + dt) = p(t) - \frac{Q_o dt}{\rho(t)V(t)}\left[\frac{1}{\beta_r} + \frac{1}{\rho(t)}\frac{\partial\rho}{\partial p}(t)\right]^{-1} \tag{9.65}$$

$$V(t + dt) = V(t)\left(1 + \frac{p(t + dt) - p(t)}{\beta_r}\right). \tag{9.66}$$

Note that there is an order to follow to solve these two differential equations, that is, the estimation of $V(t + dt)$ requires an estimation for $p(t + dt)$, so that pressure needs to be solved first in the time integration loop. In the equation for pressure (mass conservation equation), the terms related to density or its derivative with pressure are estimated with (Eq. 9.62) and (Eq. 9.63). Overall, the pseudo-code for the magma chamber FE model can be summarized as

```
% Initial conditions
V=V0;
p=p0;
N=N0;
x=x0;
% End initial conditions
dt=...;          % Choice of time step
...
for time=1:maxtime-1              % Marching equation
    p(time+1)= ...               % from Eq. 9.65
    V(time+1)= ...
    n(time+1)= ...                    % from Eq. 9.61
    r(time+1)= ...                     % from Eq. 9.62
    drdp(time+1)= ...              % from Eq. 9.63
end
```

There is an important issue to raise in the model as it is presented here. Depending on the choice for total mass fraction of water in the system N and the pressure evolution in the chamber it is possible that $n(t) < 0$, which is not realistic. Therefore a decent code should include a test to verify that $n(t) \geq 0$ at all times. An example of outputs from the model is provided in Figure 9.6. It is compared with the analytical expression published by Huppert and Woods for the case with no magma recharge and an assumed constant magma density

$$\Delta p(t) = \Delta p(0) \exp\left(-\frac{\lambda * \hat{\beta} t}{V\rho}\right), \tag{9.67}$$

with $\hat{\beta} = (1/\beta_r + \partial\rho/\partial p/\rho)^{-1}$. The agreement between numerical and analytical solutions for the temporal evolution of the overpressure is excellent.

9.3.2 Example 6: The Brusselator reaction

This model is a famous example of nonlinear autocatalytic reactions that lead to spontaneous order. It was developed by Prigogine and co-workers in Belgium [20]. The idea reposes on a system of two coupled nonlinear ODEs, which model a relatively simple auto-catalytic set of reactions. Assume

$$A \rightarrow X, \quad \text{reaction rate } k_1$$

$$2X + Y \rightarrow 3X, \quad \text{reaction rate } k_2$$

$$B + X \rightarrow Y + D, \quad \text{reaction rate } k_3$$

$$X \rightarrow E, \quad \text{reaction rate } k_4.$$

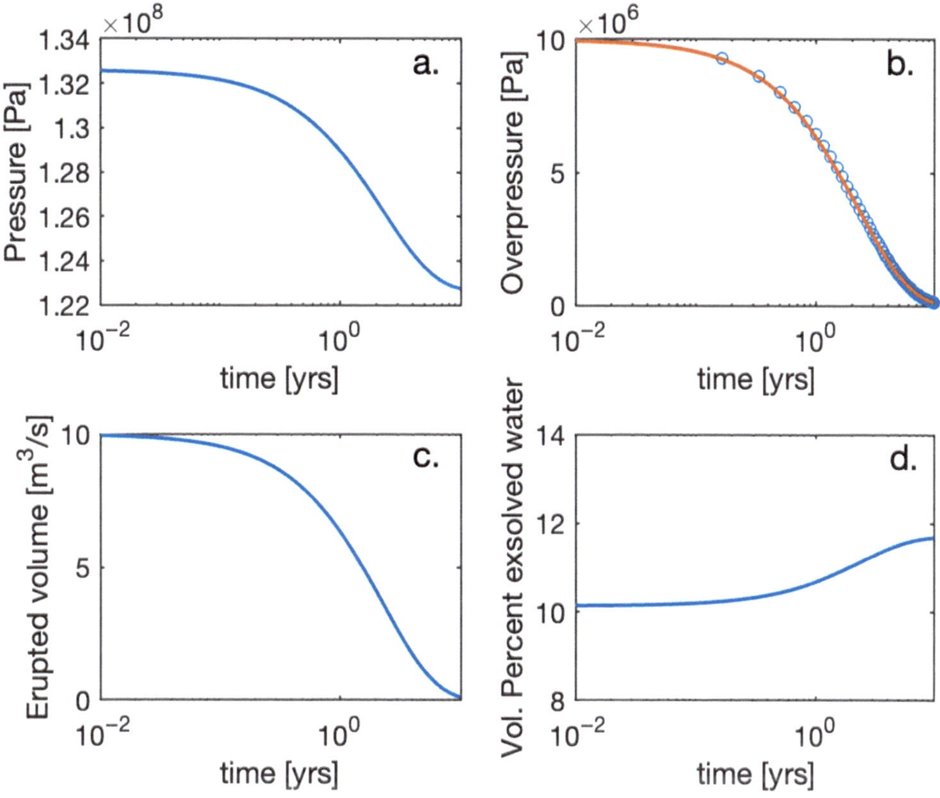

Figure 9.6 Illustration of the outputs from the forward Euler magma chamber model. a): The magma reservoir is initially in a state of overpressure $\Delta p = p_0 - p_{\text{litho}} = 20$ MPa and decreases because of $Q_o > 0$. b): Evolution of Δp with time. The numerical solution (solid line) is compared with the analytical solution defined in (Eq. 9.67). c): The mass rate of magma removed from the chamber (erupted). d): The evolution of the volume fraction of water bubbles in the chamber. The details of the setup for the calculations are given in the exercise section.

In these sets of reactions, components A and B are in vast excess so that their concentration is assumed not to change during the reaction of the byproducts X and Y. The concentration of intermediate products X and Y can be cast in differential form as

$$\frac{dX}{dt} = k_1A + k_2X^2Y - k_3BX - k_4X$$
$$\frac{dY}{dt} = -k_2X^2Y + k_3BX. \tag{9.68}$$

For simplicity, we assume all rate constants to be equal to unity. This leads to the set of two coupled nonlinear differential equations

$$\frac{dX}{dt} = A + X^2Y - BX - X \tag{9.69}$$

$$\frac{dY}{dt} = -X^2Y + BX. \tag{9.70}$$

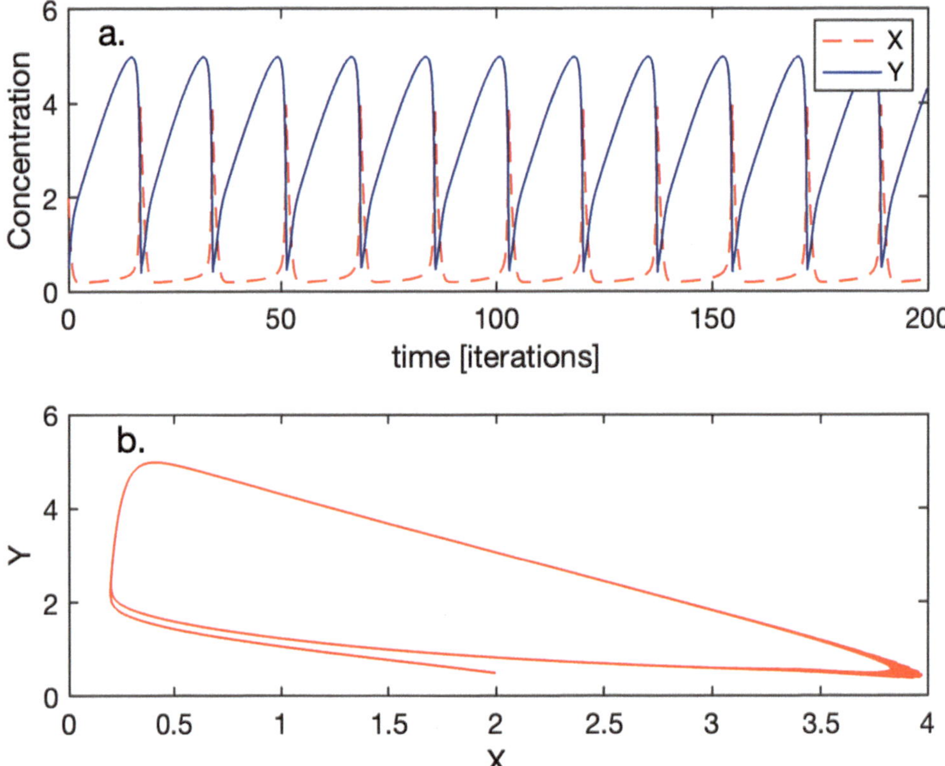

Figure 9.7 a): Temporal evolution of the concentration of both components X and Y. Note the cyclicity of the reactions. b): Phase space representation of the temporal evolution. The trajectories define an orbit in phase space, each cycle overlapping on the previous ones.

We can define a critical parameter that controls the behavior of the solution (X,Y): $\Pi = B/(1 + A^2)$. The best way to look at dynamical systems such as this one is to plot the trajectories of the solution of (X,Y) over time on an X–Y plot. One such example, for $\Pi = 1.6$, is shown in Figure 9.7

9.3.3 Exercises

1. Implement an FE approximation to the two coupled ODEs for X(t) and Y(t) defined by Equations 9.69 and 9.70. Use a time step $\Delta t = 0.1$ and reaction rates $k_1 = \ldots = k_4 = 1$.
2. Reproduce the result shown in Figure 9.7 with $\Pi = 1.6$, here $A = 0.5$, $B = 2$.
3. Now set $\Pi = 1$ with $A = 0.5$ and $B = 1.25$. Discuss your results.
4. Finally set $\Pi = 0.72$ with $A = 0.5$ and $B = 0.9$ and discuss the differences with the two previous runs.

9.3.4 Example 7: The Daisyworld

The Daisyworld model was developed by Watson and Lovelock [26] to illustrate feedbacks between biota and surface environment on the Earth. The main idea of the model is to provide a simple and idealized model to the Gaia hypothesis, that is, the idea that biota and their environment self-regulate to promote conditions amenable to life on Earth. The original Daisyworld model assumes that the Earth's surface is partitioned into three surface types: barren ground, white daisies, and black daisies. The population of the white and black daisies follows a birth/death model that is nonlinear (in terms of growth = birth). The growth of each daisy population is therefore in competition for both space and energy intake from the Sun. The growth rate of each types of daisies influences the overall albedo of the planet (effect on surface temperature) and is also a nonlinear function of surface temperature (reverse feedback). The overall model consists of two governing ODEs and six associated closure equations that describe the different feedbacks. We first review the different equations that constitute the model before discussing its implementation. Our first equation states that the Earth's surface can only be of three types: barren, white, and black daisies. Denoting the fraction of the Earth's surface covered by $x = g, w, b$, respectively barren, white, and black daisies by C_x, the trivial surface conservation becomes

$$1 = C_g(t) + C_w(t) + C_b(t). \tag{9.71}$$

If the surface coverage is known, then the overall albedo of the planet A can be computed as a weighted average

$$A(t) = C_g(t)A_g + C_w(t)A_w + C_b(t)A_b, \tag{9.72}$$

where A_x is the albedo of each surface type (fixed for the simulations). Now moving to the two governing equations that describe the temporal evolution of C_w and C_b, we define the death rate of each daisy type as a positive rate constant D multiplied by the surface coverage of that daisy type

$$\frac{dC_x}{dt} = Growth - DC_x, \quad x = w, b. \tag{9.73}$$

The growth term is assumed to be depending on the surface coverage of the type of daisies considered and the fraction of barren land available (no more growth if all the surface is covered with daisies). This leads to

$$\frac{dC_b}{dt} = C_b\left[b_b(1 - C_w(t) - C_b(t)) - D\right] \tag{9.74}$$

$$\frac{dC_w}{dt} = C_w\left[b_w(1 - C_w(t) - C_b(t)) - D\right], \tag{9.75}$$

where b_x is a birth rate that depends on temperature and daisy type.

The temperature of the atmosphere of the planet T_A is computed from a black body approximation, balancing the amount of energy received from the star (Sun) with the amount of energy re-emitted by the body (depends on overall albedo A), such that

$$T_A(t) = \left[\frac{LS_0(1 - A(t))}{4\sigma}\right]^{1/4}, \tag{9.76}$$

with σ the Stefan–Boltzmann constant (5.67×10^{-8} W/m^{2}K^4), S_0 the incoming solar energy flux (1380W/m^2 for Earth), and L a dimensionless luminosity factor for the star (= 1 for the Sun).

Assuming a very idealized atmospheric green gas effect, the surface temperature can be warmer than the atmosphere by a fixed factor here

$$T_s^4(t) = 2T_A^4(t). \tag{9.77}$$

Given each surface has its own local albedo, we can consider that the surface temperature of the planet is locally not homogeneous, and we can define a surface-type temperature for each region

$$T_i^4(t) = (1-K)\frac{LS_0}{4\sigma}(A(t) - A_i) + T_s^4(t) \quad i = w, b, g, \tag{9.78}$$

where K is a dimensionless parameter that would be 1 if the surface temperature was assumed homogeneous and 0 if perfectly heterogeneous. $K = 0.6$ is often chosen as a compromise. Finally, the birth rate constant for each daisy type is given by

$$b_i(t) = 1 - b\,(T_0 - T_i(t))^2, \tag{9.79}$$

where b is a constant that marks the sensitivity of daisies to temperature ($b = 3.265 \times 10^{-3}$ 1/K s) and T_0 the temperature of peak growth $T_0 = 295.7$ K).

It is important to pause a second to note that the governing equations are not linear because of the growth rate. One can make that a bit more obvious by rewriting (Eq. 9.74) as

$$\frac{dC_b}{dt} = b_b\left[C_b(t) - C_b^2(t) - C_w(t)C_b(t)\right] - DC_b(t) \tag{9.80}$$

$$\frac{dC_w}{dt} = b_w\left[C_w(t) - C_w^2(t) - C_w(t)C_b(t)\right] - DC_w(t). \tag{9.81}$$

This has important implications for the selection of the proper finite difference scheme. Here we focus on first-order schemes (truncation error scales linearly with the time step Δt). The BE has proven so far to be more stable than the FE discretization, so we first attempt to approach the two differential equations above from that perspective. The two ODEs being symmetrical, we only consider one for the discussion here and write the associated FDE for BE

$$\frac{C_b(t+\Delta t) - C_b(t)}{\Delta t} = b_b\left[C_b(t+\Delta t) - C_b^2(t+\Delta t) - C_w(t+\Delta t)C_b(t+\Delta t)\right] - DC_b(t+\Delta t). \tag{9.82}$$

Reorganizing the equation so that the unknowns $C_x(t+\Delta t)$ are moved to the left-hand side of the equation, we get

$$C_b(t+\Delta t)\left[1 + b_bC_b(t+\Delta t) - b_b + b_bC_w(t+\Delta t)\right] = C_b(t). \tag{9.83}$$

There is one major issue that prevent us from writing algebraic equations in linear form (matrix-vector products) to invert for the solution ($C_b(t+\Delta t), C_w(t+\Delta t)$), the nonlinearity of the equations, which is present in two separate terms $C_b^2(t+\Delta t)$ and $C_b(t+\Delta t)C_w(t+\Delta t)$, prevents us from writing a system of linear algebraic equations that can be solved by a matrix inversion. Nonlinear differential equations are not well suited for implicit (BE here) solution procedures. We therefore revert to our second option, the FE method. The FE FDEs that model the two differential equations are

$$C_b(t + \Delta t) = C_b(t)\left[1 + \Delta t b_b\left(1 - C_b(t) - C_w(t)\right) - D\Delta t\right] \quad (9.84)$$

$$C_w(t + \Delta t) = C_w(t)\left[1 + \Delta t b_w\left(1 - C_w(t) - C_b(t)\right) - D\Delta t\right]. \quad (9.85)$$

The solution procedure can be summarized by the following sequence. After initializing the different parameters (S_0, L . . .) and the initial conditions for the surface coverage of daisies ($C_b(0), C_w(0)$), every time step adheres to the following sequence of operations:

1. Compute the barren surface fraction $C_g(t)$ from (Eq. 9.71).
2. Compute the overall albedo $A(t)$ from (Eq. 9.72).
3. Compute the atmospheric temperature of the planet from (Eq. 9.76).
4. Compute the surface temperature from (Eq. 9.77).
5. Compute the surface-type dependent temperature T_i for white and black daisies from (Eq. 9.78).
6. Get the birth rate constants b_w and b_b for white and black daisies from (Eq. 9.79).
7. Solve the two marching FDEs Eq. 9.84 and Eq. 9.85 to get $C_b(t + \Delta t)$ and $C_w(t + \Delta t)$.
8. Increment time from $t \to t + \Delta t$ and start over.

The first-order approximation of the time derivative (FE) can limit the choice of time step used in the simulations. In this particular case it is not too critical, but we learn in the following sections to implement higher-order approximations for the time derivatives (Runge–Kutta methods). Figure 9.8 shows an example simulation with a luminosity $L = 1.2$ and initial surface coverage of 1% for each daisy type. The system converged to a steady-state after about 20 years,

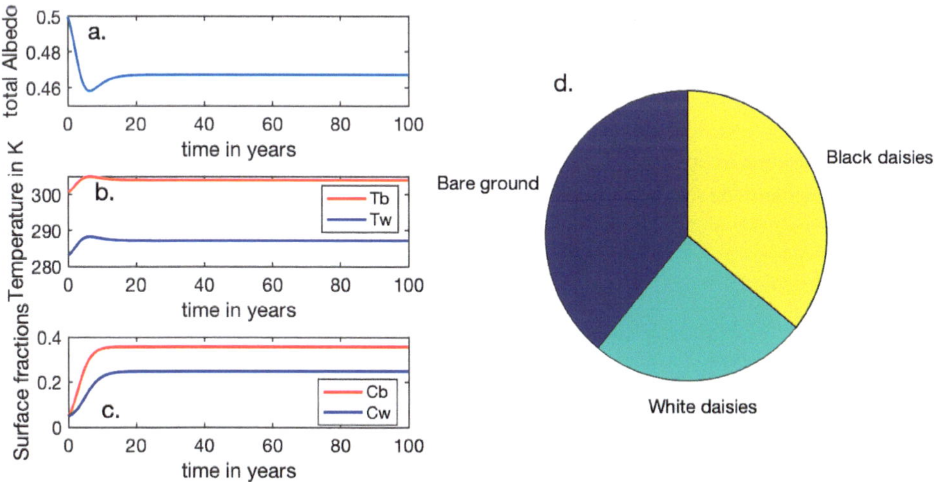

Figure 9.8 Evolution of the Daisyworld for a star luminosity of $L = 1.2$, corresponding to 20% more energy flux than the Earth receives from our Sun.

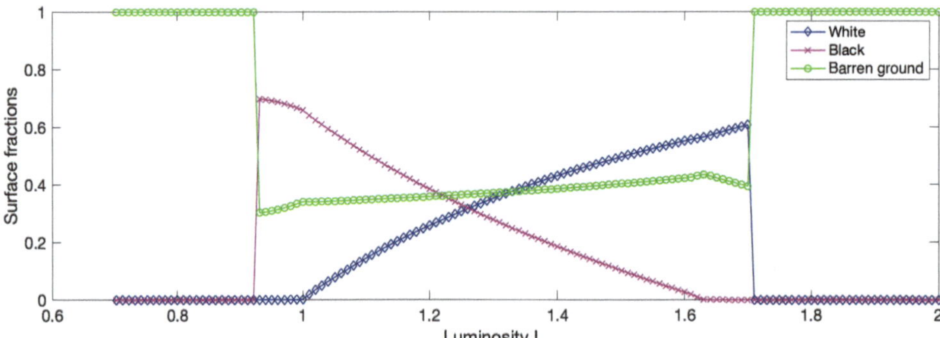

Figure 9.9 Diversity of steady-state solutions obtained for varying star luminosity (everything else remaining equal). The time integration spans 300 years to make sure all runs reach steady-state.

with black daisies covering up about 40% of the surface and white daisies slightly more than 30%. By playing with the luminosity of the stars it is possible to observe a wide range of possible solutions, from barren ground at very low luminosity to extensive daisy coverage and back to barren at high luminosity (see Figure 9.9). These sharp variations in the steady-state of the system with respect to the amount of incoming energy illustrate the nonlinear behavior of the model and the existence of feedbacks between daisy growth and changes of albedo, which consequently affect surface temperatures and birth rates of daisies.

9.3.5 **Exercises**

1. Write a script to solve the Daisyworld model described with an FE time discretization.
2. Set $C_b(0) = C_w(0) = 0.01$ as initial conditions, with a luminosity $L = 1$ (Sun). Use a run duration that covers 100 Earth years with a time step $\Delta t = 0.5$ year. Select $D = 0.3$, $A_w = 0.75$, $A_b = 0.25$, $A_g = 0.5$, $K = 0.6$, $\sigma = 5.6 \times 10^{-8}$, $T_0 = 295.7$, $b = 3.265 \times 10^{-3}$, and $S_0 = 1380$.
3. Run another simulation with $L = 1.2$, all else remaining identical to the previous run. What changes do you observe? Can you find a qualitative explanation for the difference in surface coverage?
4. Modify your script to run a sequence of Daisyworld simulations with varying luminosity values and plot the steady-state surface coverage for each run as a function of the luminosity (extend the runs to 300 years each to make sure a steady-state is reached for all conditions). Can you explain the features you observe in the graph?

Chapter 10

Higher-order ODEs

Let us start with an example to illustrate the main challenges that arise when dealing with higher-order derivatives in ordinary differential equations (ODEs). Considering the law for a generic net force F applied on a mass m, assuming, for the sake of the argument, a one-dimensional problem

$$m\frac{d^2x}{dt^2} = F, \tag{10.1}$$

So far, we have not discussed formally how to discretize derivatives of an order higher than one. Again, the mathematical tool that we use to discretize a differential equation is the Taylor series expansion, which relates differences to derivatives. We consider two distinct Taylor expansion series to approximate the position x at different times, both anchored at the reference time t

$$x(t+\Delta t) = x(t) + \frac{dx}{dt}\Delta t + \frac{d^2x}{dt^2}\frac{\Delta t^2}{2} + \frac{d^3x}{dt^3}\frac{\Delta t^3}{6} + \mathcal{O}(\Delta t^4) \tag{10.2}$$

$$x(t-\Delta t) = x(t) - \frac{dx}{dt}\Delta t + \frac{d^2x}{dt^2}\frac{\Delta t^2}{2} - \frac{d^3x}{dt^3}\frac{\Delta t^3}{6} + \mathcal{O}(\Delta t^4), \tag{10.3}$$

We can sum these two equations to get rid of all the odd order of Δt on the right-hand side

$$x(t+\Delta t) + x(t-\Delta t) = 2x(t) + \frac{d^2x}{dt^2}\Delta t^2 + \mathcal{O}(\Delta t^4), \tag{10.4}$$

Reorganizing this expression we get a second-order finite difference approximation for the second derivative

$$\frac{x(t+\Delta t) - 2x(t) + x(t-\Delta t)}{\Delta t^2} = \frac{d^2x}{dt^2} + \mathcal{O}(\Delta t^2), \tag{10.5}$$

This definition of the second-order derivative (a centered scheme, as it is symmetric around t) is very useful and we return to it many times. Using a similar logic as the method defined in Section 7.7, it is possible to generalize a procedure to get suitable discretization schemes for any order of derivatives (and any order of accuracy). However, it quickly leads to complex and large stencils that are not easy to derive nor practical to implement. There is an alternative to that approach and we introduce it in Section 10.1.

10.1 From one nth-order ODE to n first-order ODEs

The chapter introduction identifies the main challenge with discretizing high-order derivatives, the stencil (instances of the function estimated at different times) that is required to define a high-order scheme becomes rapidly complicated and heavy. It is important to show a general strategy for dealing with ODEs of order $n > 1$ that would avoid this issue.

Introduction to Numerical Modeling in Earth and Planetary Sciences, Christian Huber, Oxford University Press.
© Christian Huber (2025). DOI: 10.1093/oso/9780198802716.003.0011

We first introduce a new notation. We write the derivatives of $x(t)$ as

$$x^{(1)} = \frac{dx}{dt}, \quad (10.6)$$

$$x^{(2)} = \frac{d^2x}{dt^2}, \quad (10.7)$$

$$\ldots = \ldots \quad (10.8)$$

$$x^{(n)} = \frac{d^nx}{dt^n}, \quad (10.9)$$

As discussed, in the context of the second derivative, we define a discretization scheme for a second- or higher-order derivative by adding points to the stencil, which can become problematic and prohibitive when considering high-order derivatives and schemes with high-order accuracy (order of the truncation error). There is an alternative to this approach, and it consists in transforming a nth-order ODE into n first-order ODEs. In full generality, we define a generic nth-order ODE

$$x^{(n)} = f(t, x, x^{(1)}, \ldots, x^{(n-1)}), \quad (10.10)$$

We introduce a new set of variables $u_1, \ldots u_n$ that are function of the variable t and are defined as

$$u_1 = x, \quad (10.11)$$

$$u_2 = x^{(1)}, \quad (10.12)$$

$$\ldots = \ldots \quad (10.13)$$

$$u_n = x^{(n-1)}, \quad (10.14)$$

Now, if we differentiate each variable, we get

$$u_1^{(1)} = u_2 \quad (10.15)$$

$$u_2^{(1)} = u_3 \quad (10.16)$$

$$\ldots = \ldots \quad (10.17)$$

$$u_n^{(1)} = f(t, x, x^{(1)}, \ldots, x^{(n-1)}) = f(t, u_1, , \ldots, u_n), \quad (10.18)$$

which is now a set of n coupled first-order ODEs, the latter being the original target ODE. This substitution will make sense with the example in Section 10.2.

10.2 Example, integration of Newton's second law

Newton's second law states that the acceleration of an object is controlled by the net force applied to it and its mass

$$m\frac{d^2x}{dt^2} = F, \quad (10.19)$$

where F is the net force applied to the object of mass m and x is its position (can be a scalar, i.e., one-dimensional or a vector, it does not matter here). Decomposing this second-order ODE into two coupled first-order ODEs leads to

$$u_1 = x \tag{10.20}$$

$$u_2 = \frac{dx}{dt} = v \tag{10.21}$$

$$u_3 = \frac{d^2x}{dt^2} = \frac{dv}{dt} = \frac{F}{m}, \tag{10.22}$$

So finally the two first-order ODEs to solve are

$$\frac{dx}{dt} = v \tag{10.23}$$

$$\frac{dv}{dt} = \frac{F}{m}, \tag{10.24}$$

These two first-order equations can be solved with any scheme we have already discussed.

10.2.1 Kepler's orbit

One example of integration of Newton's second law is the motion of a planet of mass m under the gravitational force of their star (of mass M). We neglect both dissipation (tides), angular momentum conservation and interactions (force) with other planetary bodies here. In two dimensions Newton's law reads

$$m\frac{d^2\mathbf{x}}{dt^2} = \mathbf{F}_g, \tag{10.25}$$

where $\mathbf{x} = (x(t), y(t))$ is the vector of position of the planet of mass m (center of reference frame here is taken as the center of the star), the gravitational force is

$$\mathbf{F}_g = \begin{pmatrix} -\frac{GMmx(t)}{(x^2(t)+y^2(t))^{3/2}} \\ -\frac{GMmy(t)}{(x^2(t)+y^2(t))^{3/2}} \end{pmatrix}, \tag{10.26}$$

and G is the gravitational constant. Because of the two-dimensional nature of the problem (in the orbital plane), we finally have four first-order ODEs to solve, two for the velocity of the planet, here discretized,

$$\frac{v_x(t+\Delta t) - v_x(t)}{\Delta t} = -\frac{GMx(t)}{(x^2(t)+y^2(t))^{3/2}} \tag{10.27}$$

$$\frac{v_y(t+\Delta t) - v_y(t)}{\Delta t} = -\frac{GMy(t)}{(x^2(t)+y^2(t))^{3/2}}, \tag{10.28}$$

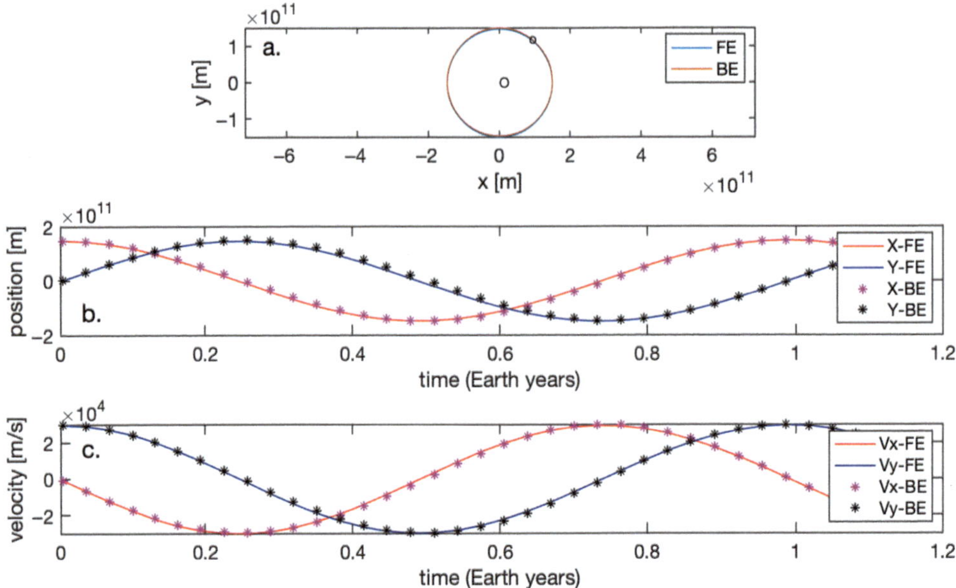

Figure 10.1 Simulation of Earth's orbit with forward and backward Euler. The time step in these simulations is $\Delta t = 10^5$ s.

and two for updating the position of the planet from the new velocity

$$\frac{x(t+\Delta t) - x(t)}{\Delta t} = v_x(t+\Delta t) \tag{10.29}$$

$$\frac{y(t+\Delta t) - y(t)}{\Delta t} = v_y(t+\Delta t), \tag{10.30}$$

Note that the scheme we wrote here is an hybrid between forward Euler (FE) and backward Euler (BE), the integration of the velocity using an FE discretization and the position a BE approach. Figure 10.1 shows an example using either FE or BE integration schemes to compute the motion of the Earth around our Sun, the time step selected here is $\Delta = 10^5$ s (about one day).

10.2.2 Drift of a charged particle in an electromagnetic field

A very similar exercise consists of the simulation of the trajectory of a charged particle in an electromagnetic field. The equations of motion for the planet remain unchanged, and we assume a two-dimensional motion by imposing the magnetic field to be perpendicular to the plane where the motion of the particle takes place. The magnetic field (fixed) is chosen so that $\mathbf{B} = (0, 0, B_z)$, and the electric field is $\mathbf{E} = (E_x, E_y, 0)$. The net force is

$$\mathbf{F} = q(\mathbf{E} + \mathbf{v} \times \mathbf{B}), \tag{10.31}$$

where $\mathbf{v}$ is the velocity of the particle and q its charge. As an example that is part of the exercises in Section 10.3, we can use a BE model to compute the trajectories of particles of different masses

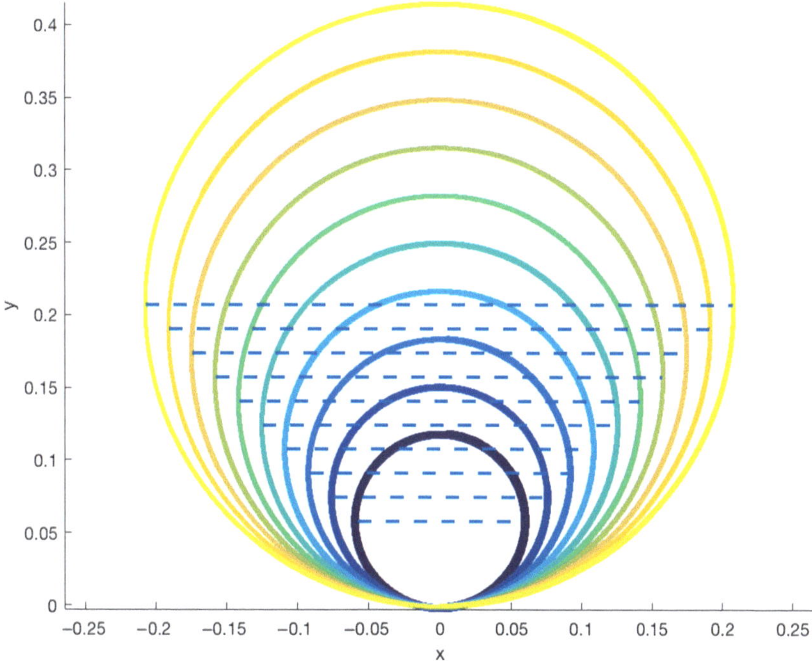

Figure 10.2 Simulation of particle trajectories in a magnetic field (initial velocity perpendicular to the fixed magnetic field). The electric field is null, which leads to circular particle trajectories. The radius of each trajectory is compared to the gyroradius (dashed line) computed with the theoretical Eq. 10.32. The time step in these simulations is $\Delta t = 10^{-10}$ s.

moving perpendicular to a fixed magnetic field. The trajectories should be circular with a radius (gyroradius) given by

$$r_m = \frac{mV_0}{|q|B_z}, \tag{10.32}$$

where m is the mass of the particle, V_0 its initial velocity, and $|q|$ the absolute value of its charge. Figure 10.2 shows an example of such simulations with 10 particles with masses ranging from 2 to 10 a.m.u. and the charge of an electron. The dashed horizontal lines compare the gyroradius predicted from theory (Eq. 10.32) with the trajectories computed numerically.

10.3 **Exercises**

1. Solve the trajectory of the charged particle following the scheme described above for the Earth's rotation around the Sun. Select $\mathbf{E} = (1, 0.5, 0)$, and $\mathbf{B} = (0, 0, 10^{-2})$, an initial velocity of $\mathbf{v} = (2 \times 10^2, 0, 0)$, the charge and mass of a proton. Use a time step $\Delta t = 10^{-8}$ seconds. Compare Forward and Backward Euler approximations for the time integration. See Figure 10.3, which provides an output from this simulation as a test for your runs.
2. Write a script that plots the trajectory of a moving particle in a magnetic field (perpendicular to the plane where the particle moves) and compare the trajectories to theoretical predictions

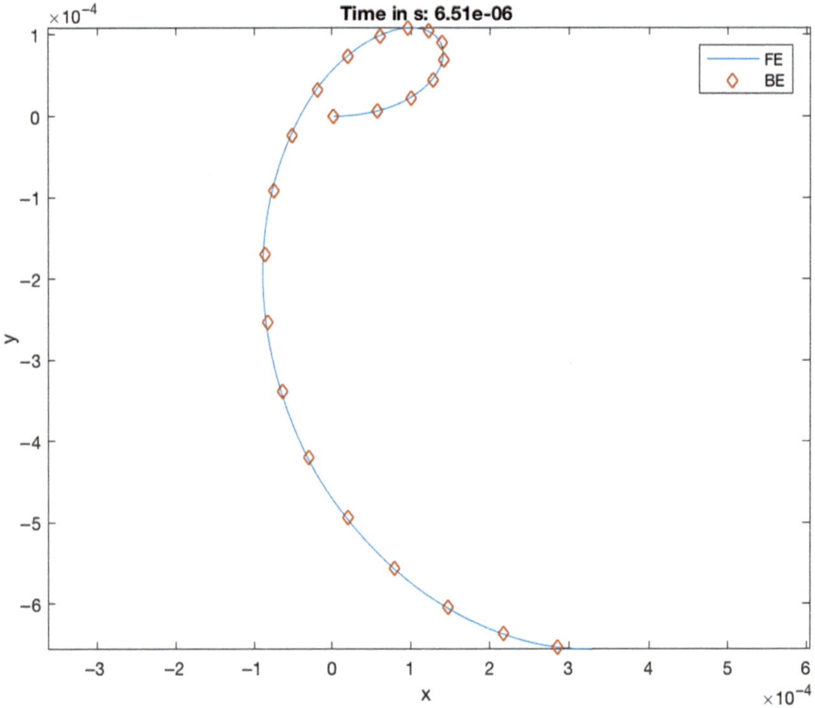

Figure 10.3 The trajectory of a charged particle in an electro-magnetic field (see Exercise 1).

of the gyroradius of the particle. Use an initial velocity $\mathbf{v} = (2 \times 10^6, 0, 0)$ and $\mathbf{B} = (0, 0, 1)$ with a time step of $\Delta t = 10^{-10}$ s.

Chapter 11

Higher-order discretization methods

There is a wealth of numerical methods that allows for higher-order accuracy when solving ordinary differential equations (ODEs). It is important here to repeat that the order of the method is not necessarily tied to the order of the differential equation. While solving a differential equation, it may be worth the effort to go beyond the first-order approximations for the derivative offered by both forward Euler (FE) and backward Euler (BE) methods. We focus here on two specific methods for ODEs. We first investigate the Verlet algorithm, which was derived for the integration of second-order ODEs associated with Newton's equations of motion. The we look at the more generic Runge–Kutta methods, which allow for high-order accuracy of first-order time derivatives.

11.1 The Verlet algorithm

The Verlet integration method was designed to solve Newton's equation of motion (second-order ODEs). There are various alternatives, but all are of the second order ($\mathcal{O}(\Delta t^2)$) and are simple to use (not much overhead compared to first-order methods). Since the 1960s Verlet integration has been used extensively to compute molecule trajectories in molecular dynamics simulations, as proposed by Loup Verlet [25].

We start by considering Newton's equation in a one-dimensional world but it can be easily extended to two- or three-dimensional spaces. Let us consider $x(t)$ to the be position of the object of interest, m its mass, and F the force applied to it. Newton's law states that

$$\frac{d^2x}{dt^2} = \frac{F}{m}. \tag{11.1}$$

Defining the acceleration $a = F/m$, we approximate first the position of the object at time $t + \Delta t$ from its previous position using a Taylor series expansion

$$x(t + \Delta t) = x(t) + \frac{dx}{dt}\Delta t + \frac{1}{2}\frac{d^2x}{dt^2}\Delta t^2 + \frac{1}{6}\frac{d^3x}{dt^3}\Delta t^3 + \ldots \tag{11.2}$$

If neither mass nor the external force depend explicitly on time, we can see that the third-order term vanishes as $\frac{d^3x}{dt^3} = 0$. We therefore consider the expansion only to the second order (Δt^2). Substituting $v = dx/dt$ and $a = d^2x/dt^2$ we get the approximation

$$x(t + \Delta t) = x(t) + v(t)\Delta t + \frac{1}{2}a(t)\Delta t^2. \tag{11.3}$$

Similarly, we can use a Taylor series expansion to approximate the new velocity (at time $t + \Delta t$) from the previous estimate of the velocity of the object

$$v(t + \Delta t) = v(t) + a(t)\Delta t, \tag{11.4}$$

Introduction to Numerical Modeling in Earth and Planetary Sciences, Christian Huber, Oxford University Press.
© Christian Huber (2025). DOI: 10.1093/oso/9780198802716.003.0012

although we mix things up a little here and consider some kind of half-step approximation instead

$$v^*(t+\Delta t) = v(t) + \frac{1}{2}a(t)\Delta t. \tag{11.5}$$

The acceleration at the new time step is retrieved from the force balance

$$a(t+\Delta t) = \frac{F(t+\Delta t)}{m}, \tag{11.6}$$

if the force depends on the position or velocity of the object (as it often does), $F(t+\Delta t) = f(x(t+\Delta t), v^*(t+\Delta t))$ can be computed from the approximations above. Finally, the updated velocity at time $t+\Delta t$ becomes

$$v(t+\Delta t) = v^*(t+\Delta t) + \frac{1}{2}a(t+\Delta t)\Delta t. \tag{11.7}$$

The solution procedure is then repeated at the next time step. In a nutshell, at every time step the following sequence of calculations is completed:

1. Approximate the new position of the object with Eq. 11.3.
2. Compute the half-step velocity of the object with Eq. 11.5.
3. Update the acceleration from the new position and velocity (half-step) of the object 11.6.
4. Get the new velocity of the object for the next iteration from Eq. 11.7.

11.2 Runge–Kutta methods

German mathematicians Carl Runge (1856–1927) and Martin Kutta (1867–1944) developed a family of multi-step time integration methods now commonly called Runge–Kutta (RK) methods. These methods are designed to solve ODEs at the required order (second or higher) by approximating the slope of the function to solve at intermediate points in the interval between t and $t+\Delta t$. These methods are easily implemented and deal very well with stiff ODEs. We define a generic first-order ODE in the following form

$$\frac{dy}{dt} = f(t,y), \tag{11.8}$$

where $y(t)$ is the solution we seek and $f(t,y)$ is the function that characterizes the ODE, in other terms, the slope of the function $y(t)$ at time t. In that notation, the FE integration yields

$$y(t+\Delta t) - y(t) = \Delta t f(t, y(t)). \tag{11.9}$$

For simplicity, we introduce the notation $K_1 = \Delta t f(t, y(t))$, which leads to a compact formulation of the FE integration

$$y(t+\Delta t) = y(t) + K_1. \tag{11.10}$$

The main idea of the RK (endpoint) method is to first conduct an FE integration of the differential equation (first step), followed by a correction of the slope dy/dt using an approximation at time $t + \alpha\Delta t$, with $0 \leq \alpha \leq 1$. Obviously, $\alpha = 0$ would not be smart, because it will fall back on the FE approximation again. The problem though is that we do not know the value of the function $f(t + \alpha\Delta t, y(t + \alpha\Delta t)$, because $y(t + \alpha\Delta t)$ is unknown. One way to overcome this issue is to approximate $y(t + \alpha\Delta t)$ from our knowledge of the first step (FE integration)

$$y(t + \alpha\Delta t) \approx y(t) + \beta K_1, \tag{11.11}$$

where both α and β remain unknown at this stage and need to be determined. Defining $K_2 = \Delta t f(t + \alpha\Delta t, y(t) + \beta K_1)$, we can now construct our two-step RK approximation of y at time $t + \Delta t$ with

$$y(t + \Delta t) = y(t) + aK_1 + bK_2, \tag{11.12}$$

where a, b are weights that also need to be determined. The main idea behind RK methods is to refine the slope approximation $f(t, y)$ by adding intermediate steps. Higher-order RK methods require more steps and therefore more weights and parameters (such as α, β) to define. Obviously, the art here lies in the determination of the values of α, β, a, and b we should use.

11.2.1 **Two-step RK**

For the two-step RK method, four parameters are introduced and need to be determined α, β, a, and b. Our objective is to derive a scheme where the truncation error is second order in Δt. For notation purposes we define here

$$y_n \equiv y(n\Delta t). \tag{11.13}$$

Using a Taylor series expansion to estimate y_{n+1} from y_n, we get, up to the third order,

$$y_{n+1} = y_n + \frac{dy}{dt}\Delta t + \frac{\Delta t^2}{2}\frac{d^2y}{dt^2} + \frac{\Delta t^3}{6}\frac{d^3y}{dt^3}\cdots \tag{11.14}$$

Using our definition of the ODE, we get

$$\frac{dy}{dt} = f(t, y(t)) \tag{11.15}$$

$$\frac{d^2y}{dt^2} = \frac{df(t, y(t))}{dt}. \tag{11.16}$$

It is important to pause here for a second and notice that $f(t, y(t))$ is a multivariate function, depending explicitly on t and $y(t)$, so its derivative (not a partial derivative here) with respect to time yields

$$\frac{d^2y}{dt^2} = \frac{df(t, y(t))}{dt} = \frac{\partial f}{\partial t} + \frac{\partial f}{\partial y}\underbrace{\frac{\partial y}{\partial t}}_{=f}. \tag{11.17}$$

Back to our Taylor series expansion in Eq. 11.14, we can now write it

$$y_{n+1} = y_n + f\Delta t + \frac{\Delta t^2}{2}\left(\frac{\partial f}{\partial t} + f\frac{\partial f}{\partial y}\right) + \mathcal{O}(\Delta t^3), \tag{11.18}$$

where f is used as a shorthand for $f(t, y(t))$. From the definition of K_2 we get

$$K_2 = \Delta t f(t_n + \alpha\Delta t, y_n + \beta K_1), \tag{11.19}$$

where $t_n = n\Delta t$ and $K_1 = \Delta t f(t_n, y_n)$. The value of $f(t_n + \alpha\Delta t, y_n + \beta K_1)$ is not known but can be approximated again with Taylor series. Here we break down the approximation into two steps for simplicity. The first step approximate change along the time variable alone is

$$f(t_n + \alpha\Delta t, y_n) = f(t_n, y_n) + \frac{\partial f}{\partial t}\alpha\Delta t + \mathcal{O}(\Delta t^2), \tag{11.20}$$

and the second step varying y in $f(t, y)$ is

$$f(t_n + \alpha\Delta t, y_n + \beta K_1) = f(t_n + \alpha\Delta t, y_n) + \frac{\partial f}{\partial y}\beta K_1 + \mathcal{O}(\Delta t^2), \tag{11.21}$$

where $\mathcal{O}(\Delta t^2)$ comes from the fact that $\beta^2 K_1^2 \sim \Delta t^2$. Combining both steps for the approximation leads to

$$f(t_n + \alpha\Delta t, y_n + \beta K_1) = f(t_n, y_n) + \frac{\partial f}{\partial t}\alpha\Delta t + \frac{\partial f}{\partial y}\beta K_1 + \mathcal{O}(\Delta t^2). \tag{11.22}$$

We can now substitute back into K_2 and write

$$K_2 = \Delta t\left[f(t_n, y_n) + \alpha\Delta t\frac{\partial f}{\partial t} + \beta \underbrace{K_1}_{=\Delta t f(t_n, y_n)} \frac{\partial f}{\partial y}\right] + \mathcal{O}(\Delta t^3). \tag{11.23}$$

Back to our definition of the two-step RK method in Eq. 11.12, using our expressions for K_1 and K_2, we get

$$y_{n+1} = y_n + a\Delta t f(t_n, y_n)(a + b) + b\Delta t^2\left(\alpha\frac{\partial f}{\partial t} + \beta f(t_n, y_n)\frac{\partial f}{\partial y}\right) + \mathcal{O}(\Delta t^3). \tag{11.24}$$

Comparing Eq. 11.24 with Eq. 11.18, we can conclude that

$$a + b = 1 \tag{11.25}$$

$$\alpha b = \beta b = \frac{1}{2}. \tag{11.26}$$

Obviously these three constraints are insufficient to solve for a, b, α, and β, but we add an arbitrary constraint $\alpha = \beta = 1$ to close the model. This fixes the weights $a = b = 1/2$. The two-step RK (endpoint) model is therefore

$$y_{n+1} = y_n + \frac{K_1 + K_2}{2}, \tag{11.27}$$

with

$$K_1 = \Delta t f(t_n, y_n) \quad \text{FE} \tag{11.28}$$

$$K_2 = \Delta t f(t_{n+1}, y_n + K_1). \tag{11.29}$$

11.2.2 **Four-step RK**

A similar approach will lead to the four-step RK (endpoint) scheme, where now the goal is to produce a time integration that has a truncation order that is $\mathcal{O}(\Delta t^5)$. We do not derive it here, but the scheme requires four different estimations of the slope of dy/dt between t and $t + \Delta t$. These four slope approximations are defined by

$$K_1 = \Delta t f(t_n, y_n) \tag{11.30}$$

$$K_2 = \Delta t f\left(t_n + \frac{\Delta t}{2}, y_n + \frac{K_1}{2}\right) \tag{11.31}$$

$$K_3 = \Delta t f\left(t_n + \frac{\Delta t}{2}, y_n + \frac{K_2}{2}\right) \tag{11.32}$$

$$K_4 = \Delta t f(t_n + \Delta t, y_n K_3). \tag{11.33}$$

The approximation of the function $y(t)$ at time t_{n+1} becomes then

$$y_{n+1} = y_n + \frac{K_1 + 2K_2 + 2K_3 + K_4}{6}. \tag{11.34}$$

11.2.3 **Back to Kepler's orbit model**

It is worth illustrating how RK methods can be implemented in the context of an application. We revisit the solution of planetary orbits from Newton's law of motion. Using the same approach as in Section 10.2.1, in a Cartesian (two-dimensional) reference frame, we can write the equations of motion as two second-order ODEs that can be decomposed into four first order ODEs

$$\frac{dV_x}{dt} = \frac{F_{gx}}{m}, \tag{11.35}$$

$$\frac{dV_y}{dt} = \frac{F_{gy}}{m}, \tag{11.36}$$

$$\frac{dx}{dt} = V_x, \tag{11.37}$$

$$\frac{dy}{dt} = V_y. \tag{11.38}$$

Instead of using the FE or BE schemes as before, we can now integrate each of these first-order ODEs with a two-step RK solver. It would be more accurate for a given choice of time step Δt, but it comes with a price, as the solution procedure needs to follow a prescribed order. The reason for a specific sequence for the solution procedure is because the two components of the gravitational force depend on the planet's position, but not on its velocity. At first we can define the first step of the RK approach for each of the four ODEs above as

$$K_{1,Vx} = \frac{\Delta t}{m} F_{gx}(t, x(t), y(t)) \tag{11.39}$$

$$K_{1,Vy} = \frac{\Delta t}{m} F_{gy}(t, x(t), y(t)) \tag{11.40}$$

$$K_{1,x} = \Delta t V_x(t, x(t), y(t)) \tag{11.41}$$

$$K_{1,y} = \Delta t V_y(t, x(t), y(t)), \tag{11.42}$$

and so far we can solve for any K_1 in any sequence as the right-hand sides only depend on information from the previous (known) step (time t). In other words, these first steps of integration are equivalent to an FE approach. It is the second step of the RK method that requires more care. Because the gravitational force depends only on the planet's position, we can proceed to the computation of the second step for the velocity field first

$$K_{2,Vx} = \frac{\Delta t}{m} F_{gx}(t + \Delta t, x(t) + K_{1,x}, y(t) + K_{1,y}) \tag{11.43}$$

$$K_{2,Vy} = \frac{\Delta t}{m} F_{gy}(t + \Delta t, x(t) + K_{1,x}, y(t) + K_{1,y}). \tag{11.44}$$

We can now integrate the velocity field from t to $t + \Delta t$

$$V_x(t + \Delta t) = V_x(t) + \frac{K_{1,Vx} + K_{2,Vx}}{2} \tag{11.45}$$

$$V_y(t + \Delta t) = V_y(t) + \frac{K_{1,Vy} + K_{2,Vy}}{2}. \tag{11.46}$$

Then we can update the position of the planet to

$$x(t + \Delta t) = x(t) + \frac{K_{1,x} + K_{2,x}}{2} \tag{11.47}$$

$$y(t + \Delta t) = y(t) + \frac{K_{1,y} + K_{2,y}}{2}, \tag{11.48}$$

with

$$K_{2,x} = \Delta t V_x(t + \Delta t) \tag{11.49}$$

$$K_{2,y} = \Delta t V_y(t + \Delta t). \tag{11.50}$$

An example simulation that compares the solution from the two-step RK solver with those of the FE and BE (first-order) schemes is shown in Figure 11.1.

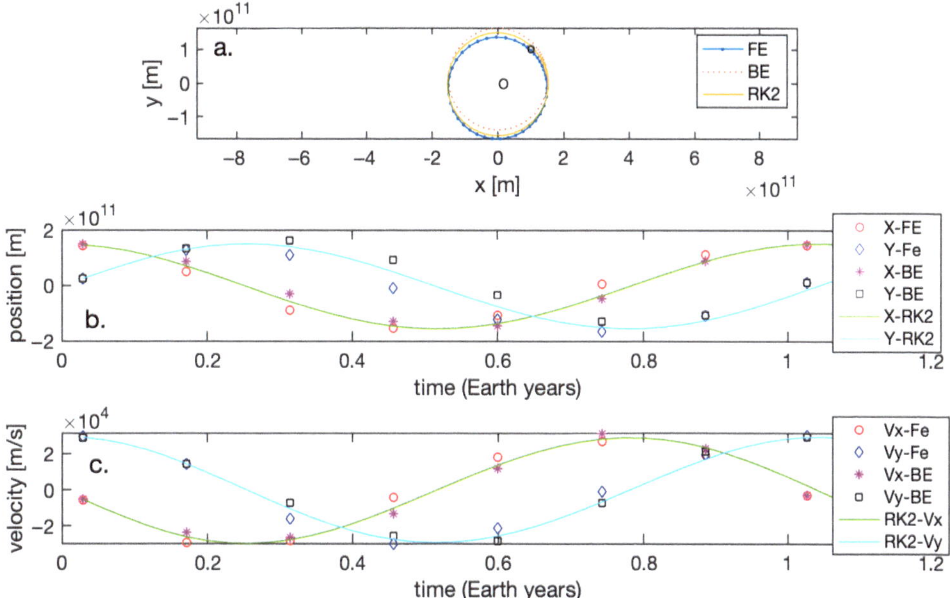

Figure 11.1 Simulation of Earth's orbit with different schemes. The choice of time step here is $\Delta t = 9 \times 10^5$ s. Note the discrepancy between forward and backward Euler schemes (compared to Figure 10.1 with $\Delta t = 10^5$ s). The two-step Runge–Kutta solver offers a more accurate solution.

11.3 **Exercises**

1. Compare the performance of the FE and BE with the two-step RK solver for the orbit of Earth with time steps ranging from $10^5 \leq \Delta t \leq 10^6$ s.

Part III

Numerical modeling, partial differential equations (PDEs)

Chapter 12

Important mathematical notions when working with partial differential equations (PDEs)

The concept of truncation error, that is, the error we introduce by discretizing derivatives, remains relevant when considering partial differential equations (PDEs). We now consider how in multivariate functions and their partial derivatives (with respect to different variables) the truncation error not only includes truncation terms for time derivatives, but also from spatial derivatives. Fortunately, Taylor series expansions easily generalize to multivariate functions and partial derivatives (see Part I). Importantly, the truncation error may (and most often will) have a different order with respect to the different variables considered. For instance, a finite difference scheme can have a truncation error that is first order in time or $\mathcal{O}(\Delta t)$ and perhaps second order in space or $\mathcal{O}(\Delta x^2)$.

When considering truncation errors of numerical schemes designed to solve equations that contain derivatives with respect to time and space, we see that some of the leading-order terms of the truncation error can play some nasty tricks (see (Eq. 12.2)). More specifically, in particular cases, we can learn a lot about numerical stability just by looking at the expression of some of the leading terms of the truncation error. Truncated terms introduce spurious effects that can affect (and in some cases destabilize) the numerical solution. These spurious effects are not limited to, but generally come in, two flavors: **numerical dispersion** and **numerical diffusion**. We begin with an introduction of diffusion and dispersion behaviors and then follow with a discussion of their expression as numerical artifacts in our solution.

12.1 Numerical dispersion

We start with a discussion of what dispersion means, before reframing it within the context of truncation errors. We resort to some simple concepts from complex calculus. The fundamental concepts that we use are explained in greater detail in Part I and readers are encouraged to review them if necessary. Dispersion is a feature associated with the propagation of waves in a medium. Imagine someone shouts and the sound emitted contains a superposition of waves within a range of frequencies $\omega_1 \leq \omega \leq \omega_n$. If waves at different frequencies travel at different velocities, the pitch, as it propagates from its source to a distant receiver, gets distorted. Over a greater time interval, the further one goes and the worse the situation becomes as the time difference at which waves with different frequencies arrive at the receiver increases. In a nutshell, this effect is what we refer to wave dispersion. In a more mathematical framework, if a wave with frequency ω_k for $k' \neq k$, travels at a speed c_k ($c_k = \omega_k/k$ is generally called the **phase velocity**) and $c_k \neq c_{k'}$ when $\omega_k \neq \omega_{k'}$, then

Introduction to Numerical Modeling in Earth and Planetary Sciences, Christian Huber, Oxford University Press.
© Christian Huber (2025). DOI: 10.1093/oso/9780198802716.003.0013

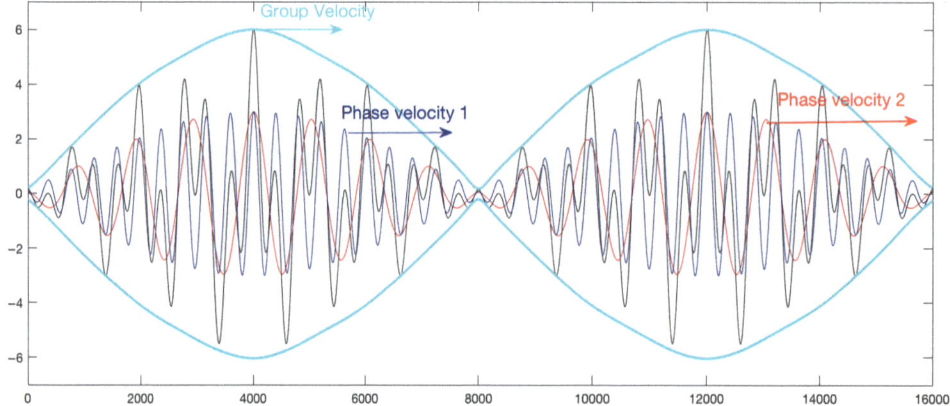

Figure 12.1 Illustration of the concept of phase and group velocity for waves with different frequencies. The thick solid lines (cyan) show the envelope of the wave packet composed of the superposition of two waves with different frequencies (red and blue), the superposition is shown in black. The group velocity is the velocity of the wave packet itself, while the phase velocities are the actual velocity of the individual waves.

the wave propagation is dispersive. Alternatively, if waves with any frequency travel all at the same speed, then the propagation is nondispersive. Let us define another velocity, the **group velocity**,

$$v_k = \frac{\partial \omega_k}{\partial k}, \tag{12.1}$$

which, in other terms, refers to the variation of (reciprocal time) frequency with respect to variations in wavenumber (spatial frequency). The group velocity is not the velocity of each individual wiggles ($c_k = \omega_k/k$), but it would be ideally the velocity of a wave packet or the envelope of the superposition of the waves (see Figure 12.1). In a nondispersive case, the phase velocities and group velocity are equal to each other, and then individual waves travel within the packet (envelope) at the same speed as the packet itself. So the wave packet does not change appearance (or pitch) as it travels.

We take some mathematical examples of partial differential equations out of context (some that we will see again shortly), for the sake of illustrating what a dispersive PDE refers to. Our first example is the linear advection equation, which is the first PDE we study in depth

$$\frac{\partial u(t,x)}{\partial t} = -a\frac{\partial u(t,x)}{\partial x}, \tag{12.2}$$

where a is a constant with units of velocity [*distance*/*time*]. A wave in time and space with (time) frequency ω_k and spatial frequency (wavenumber) k can be expressed as the real part of

$$u(x,t) \approx A\exp(ikx)\exp(i\omega_k t), \tag{12.3}$$

with A the amplitude (see Part I). We loosen up the notation here and often omit that we are considering the real part of a complex function, but it remains implied. Using the approximation of Eq. 12.3 we get by differentiating $u(t,x)$

$$\frac{\partial u}{\partial t} = i\omega_k u(t,x) \tag{12.4}$$

$$\frac{\partial u}{\partial x} = iku(t,x). \tag{12.5}$$

Inserting these expressions back into our PDE, we get

$$i\omega_k u(t,x) = -iaku(t,x), \tag{12.6}$$

which being true for any value of $u(t,x)$ means that

$$\omega_k = -ak. \tag{12.7}$$

In terms of phase (c_k) and group (v_k) velocities, this yields

$$c_k = \frac{\omega_k}{k} = -a \tag{12.8}$$

$$v_k = \frac{\partial \omega_k}{\partial k} = -a, \tag{12.9}$$

and therefore the phase and group velocities are identical, and the linear advection equation is nondispersive.

We now consider a different PDE

$$\frac{\partial u(t,x)}{\partial t} = -p\frac{\partial^3 u(t,x)}{\partial x^3}, \tag{12.10}$$

where p is a constant with units of $[distance^3/time]$. Repeating the same procedure with the same approximation for $u(t,x)$ (Eq. 12.3) leads to

$$\frac{\partial u}{\partial t} = i\omega_k u(t,x) \tag{12.11}$$

$$\frac{\partial^3 u}{\partial x^3} = -ik^3 u(t,x). \tag{12.12}$$

Back into the PDE, we get now

$$\omega_k = pk^3. \tag{12.13}$$

In terms of phase (c_k) and group (v_k) velocities, this yields

$$c_k = \frac{\omega_k}{k} = k^2 p \tag{12.14}$$

$$v_k = \frac{\partial \omega_k}{\partial k} = 3k^2 p, \tag{12.15}$$

and therefore $c_k \neq v_k$ because of the factor 3. This PDE is therefore dispersive.

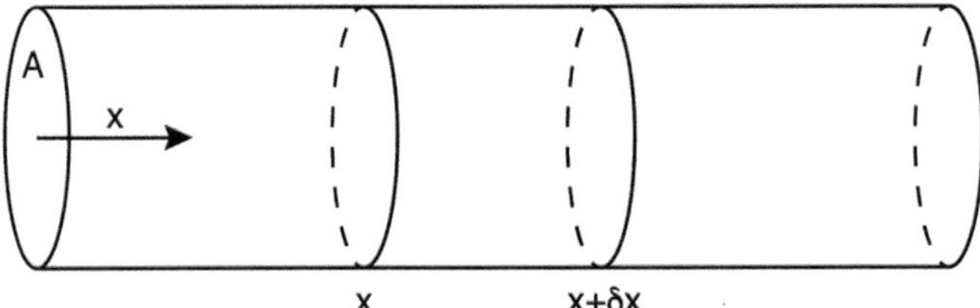

Figure 12.2 One-dimensional domain for the derivation of the diffusion equation. Imagine a long rod, insulated on the sides (radially), where heat is transferred solely along the x-axis.

We have discussed what dispersion means within the context of a PDE. This is related to the concept we introduce now: **numerical dispersion**. However, there is a subtle, yet important, difference. Numerical dispersion refers to an artifact of the discretization that introduces a spurious dispersion into an equation that may or may not be dispersive in its own right. In other words, numerical dispersion arises when terms in the truncation error introduce a dispersion effect that is not expected from the PDE itself. Because numerical dispersion originates from the truncation error, the choice of time step Δt and grid spacing Δx will strongly influence it. We provide some example of numerical dispersion in Chapter 14.

12.2 **Numerical diffusion**

Before discussing numerical diffusion it is important to get acquainted with the process of diffusion and what diffusion does to a numerical function. We first go over a simple derivation of a diffusion equation, and a more thorough derivation is provided in the section devoted to the diffusion equation. Diffusion is ubiquitous in physical sciences, that is, it comes out of any analysis deriving a mass or energy conservation equation near equilibrium, where fluxes of mass or energy are assumed to depend linearly on the gradient of the quantity of interest (see more in the diffusion equation section again). For the present, we assume a one-dimensional problem along a dimension called x. For simplicity, we assume diffusion of heat along a rod and neglect radial heat exchanges (no heat exchange between the rod and its surroundings). We first consider a small section of the rod between positions x and $x + \delta x$ and conduct a simple energy balance at two instants in time separated by an interval δt. Lastly, we assume no phase change or additional contribution to the heat balance than the conduction of heat along the rod.

Introducing the specific heat c, which measures the ability of a material to store thermal energy (i.e., how much energy is required to elevate the temperature of the body of unit mass by one unit of temperature) and the density of the material ρ, both assumed homogeneous here (i.e., no spatial dependence), the energy at time t in the section between x and $x + \delta x$ is

$$E(t) = \rho c A T\left(x + \frac{\delta x}{2}, t\right)\delta x, \tag{12.16}$$

where A is the cross section of the rod (assumed constant as well) and the average temperature is taken at the midpoint between x and $x + \delta x$, an arbitrary choice that will not influence the derivation. Now, we can repeat a similar estimate of the thermal energy in the same section of the rod at a time $t + \delta t$ and assign the difference as ΔE

$$\Delta E = \rho c A \left[T\left(x + \frac{\delta x}{2}, t + \delta t\right) - T\left(x + \frac{\delta x}{2}, t\right) \right] \delta x. \tag{12.17}$$

This change in energy must be caused by an imbalance between incoming and outgoing energy flux during this interval of time δt

$$\Delta E = A \delta t \left(q(x) - q(x + \delta x) \right). \tag{12.18}$$

Using the expression for ΔE and dividing both sides by $(A \delta x \delta t)$, we obtain

$$\frac{\rho c \left[T\left(x + \frac{\delta x}{2}, t + \delta t\right) - T\left(x + \frac{\delta x}{2}, t\right) \right]}{\delta t} = \frac{q(x) - q(x + \delta x)}{\delta x}. \tag{12.19}$$

Assuming now that $\delta x, \delta t \rightarrow 0$ and using the definition of the derivative, we get the following PDE

$$\rho c \frac{\partial T}{\partial t} = -\frac{\partial q}{\partial x}. \tag{12.20}$$

Heat is transferred by conduction, and we can use Fourier's law for heat transfer to relate the heat flux q to the local temperature gradient

$$q = -k \frac{\partial T}{\partial x}, \tag{12.21}$$

where k is the heat conductivity (units Watts/meter/Kelvin) assumed to be homogeneous as well. We finally obtain, after reorganizing terms,

$$\frac{\partial T}{\partial t} = \underbrace{\frac{k}{\rho c}}_{\kappa} \frac{\partial^2 T}{\partial x^2}, \tag{12.22}$$

where the ratio of material properties κ is referred to as the thermal diffusivity (units length^2/time).

We can now study the behavior of the diffusion equation to get a sense for what numerical diffusion entails. Using the same approach as for numerical dispersion, we introduce a wave in time and space (Eq. 12.3) into the diffusion PDE and get

$$\frac{\partial T}{\partial t} = i \omega_k T(t, x) \tag{12.23}$$

$$\frac{\partial^2 T}{\partial x^2} = -k^2 T(t, x). \tag{12.24}$$

Substituting these expressions into the diffusion equation PDE allows us to relate the wavenumber k to the angular frequency ω_k

$$i\omega_k = -k^2\kappa. \tag{12.25}$$

Going back to our wave mode description of $T(t, x)$, we get that the temperature field evolves under diffusion according to a superposition of waves such that

$$T(t, x) = \sum_k \underbrace{A_k \exp(-k^2\kappa t)}_{amplitude(t)} \exp(ikx). \tag{12.26}$$

This result, proper to the diffusion equation, is worth discussing in details. As a continuous function can always be written as an infinite sum of waves (such as (Eq. 12.3); see Part I), a diffusion process takes each of the modes and decays their initial amplitude (A_k) exponentially with time at a rate given by the square of the actual wavenumber of the mode. What does that mean? It means that: 1) diffusion decays spatial heterogeneities, and 2) that the rate at which they decay depends on the lengthscale that characterizes these heterogeneities. In other words, spatially confined and sharp peaks decay faster than large and shallow amplitude variations. This is can be made clearer by replacing the wavenumber k by its associated wavelength $\lambda_k = 2\pi/k$ so that the rate of amplitude decay over time for a mode characterized by the wavenumber k is $\exp\left[-\left(\frac{2\pi}{\lambda_k}\right)^2 \kappa t\right]$. The decay rates of two modes with wavelengths differing by an order of magnitude will end up being different by two orders of magnitude (slower for the long wavelength mode). So diffusion tends to behave as the great equalizer, that is, it smooths out heterogeneities. More accurately, as the second spatial derivative of a function $f(x)$ is related to its local curvature (see Part I, Chapter 1), the diffusion equation applied to $f(x)$ can be conceptualized as a process that tends to minimize the curvature of $f(x)$.

The second very important note to make with regards to Eq. 12.26 is what we considered so far $\kappa > 0$ in the discussion. This is warranted because of Le Chatelier's principle, which describes the fact that stable systems return to equilibrium after being pushed away by an external perturbation. This can only be true if the diffusion coefficient κ is positive (i.e., amplitude of the perturbation decays exponentially over time because of diffusion). If we were to perturb the equilibrium temperature distribution within the rod and assume a negative thermal diffusion coefficient κ, the initial perturbation would grow and not decay exponentially in time instead, leading to an unbounded amount of energy stored in the rod! Negative diffusion coefficients therefore lead to unstable and unphysical systems. They will be one of our nemeses when they arise in the truncation errors, as they result in unstable numerical schemes.

Now that we have an appreciation of diffusion, we can extend the concept to **numerical diffusion**, where a spurious diffusion behavior is introduced in the finite difference equation (FDE) solution because of a diffusive term hidden in the truncation error. For this to happen, a first derivative in space in the PDE interacts with a second derivative in space in the truncation error of the FDE. If the numerical diffusion coefficient is positive, this leads to a smearing and decay of heterogeneities, which is faster for features with smaller wavelengths. So positive numerical diffusion does not lead (in general) to numerical instability; on the contrary. Rather, it stabilizes the scheme but introduces errors that can be pronounced. However, the situation is entirely different if the diffusion coefficient of the numerical diffusion term is negative. In that case, numerical diffusion invariably leads to an unstable numerical scheme!

Chapter 13

Von Neumann stability analysis: Concepts

John von Neumann was one of the first mathematicians to see the potential of using computers to solve difficult partial differential equations (PDEs). Based on earlier work by Crank and Nicolson, he developed a rigorous approach to testing the stability of finite difference schemes for PDEs. The concept is rooted on the theory of the Fourier series and the representation of continuous functions as a sum of an infinite basis of functions that are easy to differentiate. Most frequently, we use cosine, sine expansion series, or Fourier series (see Part I for discussion of these series and the properties of these basis functions).

The overarching idea is that the target function, that is, the function we are solving for with the numerical scheme, can be written as an infinite sum of waves over space with amplitudes that vary over time. In that context, numerical stability (i.e., boundedness of the solution) is reduced to conditions on how the amplitude of each mode (wave) evolves over time. We define the concept of **absolute stability as** the condition under which every mode's amplitude shrinks or remains constant in magnitude over time. This guarantees overall stability of a scheme.

We use the von Neumann stability analysis on several schemes for diffusion, advection–diffusion, and wave equations. The advection equation uses a less formal (and cruder) analysis based on the analysis of numerical diffusion within the truncation error of the different schemes.

Overall, the von Neumann stability analysis follows these steps:

1. Write the target function that you aim to solve with the finite difference equation (FDE) as cosine, sine, or Fourier series. Assume that variables can be separated so that the coefficients or weights of each wave, here referred to as the amplitudes, vary only over time, while the basis functions (cosine, sine, or exp) carry spatial information.
2. Introduce this series expansions into the FDE and substitute the function for its expression based on amplitudes and waves.
3. Use the orthogonality of the basis functions (see Part I for more information) to reduce the infinite sums to a single component or mode.
4. From the reduced expression extract a condition for the amplitude of the selected mode not to grow with time, which provides the condition for absolute stability, given the same approach can be repeated for any selected mode.

At the end, we obtain conditions for stability in terms of transport coefficients that belong to the PDE and the choice of time step and grid spacing that is used for the FDE.

Introduction to Numerical Modeling in Earth and Planetary Sciences, Christian Huber, Oxford University Press.
© Christian Huber (2025). DOI: 10.1093/oso/9780198802716.003.0014

Chapter 14

One-dimensional advection equation

The linear advection equation for a scalar quantity $T(x, t)$ (here in one dimension) reads

$$\frac{\partial T}{\partial t} + u\frac{\partial T}{\partial x} = 0. \tag{14.1}$$

It is a linear partial differential equation (PDE) with first-order derivatives in time and space and where the scalar quantity u represents the velocity with which the quantity defined by T is transported through the domain. It naturally arises when formulating conservation statements (e.g., conservation of mass or energy). In this section, we derive this equation in an intuitive and physical perspective.

For example, the function $T(x, t)$ represents the temperature measured at time t and position x (note that the same argument is valid for different measurable quantities). Assume that as you go for a hike, you take measurements of the temperature at different times, t_1 and $t_2 > t_1$, which correspond to different locations, x_1 and x_2. Let us denote the difference in temperature recorded between these two points

$$dT \equiv T(x_2, t_2) - T(x_1, t_1). \tag{14.2}$$

Let us break the path in (x, t) space to go from (x_1, t_1) to (x_2, t_2) into two conceptual steps where only one of the two independent variables (x and t) are varied alternatively. If we decide to stay put at position x_1 during the interval of time $t_1 \rightarrow t_2$, then the temperature we would measure at t_2 can be estimated from a Taylor expansion around $T(x_1, t_1)$

$$T(x_1, t_2) = T(x_1, t_1) + \frac{\partial T}{\partial t}|_{t_1}\Delta t + \mathcal{O}\left(\Delta t^2\right), \tag{14.3}$$

where $\Delta t = t_2 - t_1$. Considering the second step in our procedure, we need to travel (at infinite speed) from position x_1 to x_2 at time t_2, where again the strategy is to consider exploring the path one variable at a time. This brings up, using Taylor expansions again,

$$T(x_2, t_2) = T(x_1, t_2) + \frac{\partial T}{\partial x}|_{t_2}\Delta x + \mathcal{O}\left(\Delta x^2\right), \tag{14.4}$$

where now we introduce $\Delta x = x_2 - x_1$. Combining these two equations yields

$$dT = \frac{\partial T}{\partial t}|_{t_1}\Delta t + \frac{\partial T}{\partial x}|_{t_2}\Delta x + \mathcal{O}\left(\Delta x^2, \Delta t^2\right). \tag{14.5}$$

Introduction to Numerical Modeling in Earth and Planetary Sciences, Christian Huber, Oxford University Press.
© Christian Huber (2025). DOI: 10.1093/oso/9780198802716.003.0015

Dividing both sides with Δt and taking the limit $\Delta t \to 0$ leads to

$$\frac{dT}{dt} = \frac{\partial T}{\partial t} + \underbrace{\lim_{\Delta t \to 0} \frac{\Delta x}{\Delta t}}_{\equiv u} \frac{\partial T}{\partial x}. \tag{14.6}$$

This equation highlights that when we travel within a field $T(x, t)$, changes in the measurement T can arise because of two independent causes: 1) locally the field T can vary temporally (in our example, temperature fluctuations recorded by a fixed thermometer), and 2) in a heterogeneous T field, changing positions and the rate at which this change happens ($u \equiv \lim_{\Delta t \to 0} \Delta x/\Delta t$) influence the rate of change of measured T.

The linear advection equation is a hyperbolic PDE. In the following sections, we introduce three different discretization schemes to solve this equation (there are many more). We selected these three schemes because they either highlight some of the challenges and pathologies that can often affect numerical solutions or because they provide a decent and widely used approximation to the differential equation.

14.1 Scheme 1: FE(time)–CE(space)

The first finite difference scheme we pay attention to is a forward Euler (FE) (time)–centered Euler (CE) (space) (FE–CE) scheme. For notation purposes (and we stick to this notation for the rest of the book), we define superscripts to reference the time step and subscripts to define the grid point (position) such that

$$T_k^n \equiv T(k\Delta x, n\Delta t). \tag{14.7}$$

With this notation, the finite difference equation (FDE) is

$$\frac{T_k^{n+1} - T_k^n}{\Delta t} + u\frac{T_{k+1}^n - T_{k-1}^n}{2\Delta x} = 0. \quad k = 1, \ldots, N. \tag{14.8}$$

It is simple to implement (it is a simple marching scheme with only one unknown T_k^{n+1} per algebraic equation) and offers a first-order (in time) but a second-order (in space) accuracy (see discussion of the truncation error in Section 14.1.3).

14.1.1 Implementation

A simple equation for each of the T_k^{n+1} can be implemented

$$T_k^{n+1} = T_k^{n+1} + \frac{C}{2}\left(T_{k+1}^n - T_{k-1}^n\right), \tag{14.9}$$

where $C = u\Delta/\Delta x$ is a dimensionless factor called the Courant number. It depends both on the velocity characterizing the advection process and the coarseness of the discretization.

```
Dt=...;
Dx=...;
u=...;
C=u*Dt/Dx;
T_old=...; % Initial condition
for time=1:maxtime-1 % Marching equation
        for position=2:N-1
                    T(position)=T_old(position)+... \\
         C/2*(T_old(position+1)-T_old(position-1));
   end
% Boundary conditions to complete T(1) and T(N)
...
end
```

Something not discussed up to this point is boundary and initial conditions. Given that the differential equation is first order in time and space, it requires one initial condition $T(x, t = 0)$ and one boundary condition at $x = 0$ (left boundary) or $x = L$ (right boundary).

14.1.2 **Performance**

The problem we solve to test the scheme is that of a rectangle wave with initial condition

$$T(x,0) = \begin{cases} 1 & \text{if} \quad l \le x \le l + \delta, \\ 0 & \textit{otherwise}. \end{cases} \tag{14.10}$$

The velocity u is set to a positive constant value and the boundary condition for this example is periodic, that is, $T_{N+1} = T_1$ and $T_0 = T_N$ for all times.

Outputs from the numerical experiments are shown in Figure 14.1 (three snapshots in time, early after the onset of the simulation). The numerical solution for this scheme does not preserve the shape of the initial wave that is expected from the PDE and involves growing oscillations originating from the edges of the initially rectangular wave. We now proceed to study the consistency and stability of the scheme.

14.1.3 **Consistency**

To establish the consistency of the scheme, we need to determine the truncation error $\mathcal{T}(\Delta x, \Delta t)$ and verify that it vanishes as the discretization in space Δx and time Δt both $\to 0$. Again, we resort to the Taylor expansion series to link the finite difference to the (partial) differential equation. We do so by considering expansion in time and space around the discrete value of the $T(x, t)$ function at $x = x_k = k\Delta t$ and $t = t_n = n\Delta t$,

$$T_k^{n+1} = T_k^n + \frac{\partial T}{\partial t}|_{t_n} \Delta t + \frac{\Delta t^2}{2} \frac{\partial^2 T}{\partial t^2}|_{t_n} + \mathcal{O}(\Delta t^3). \tag{14.11}$$

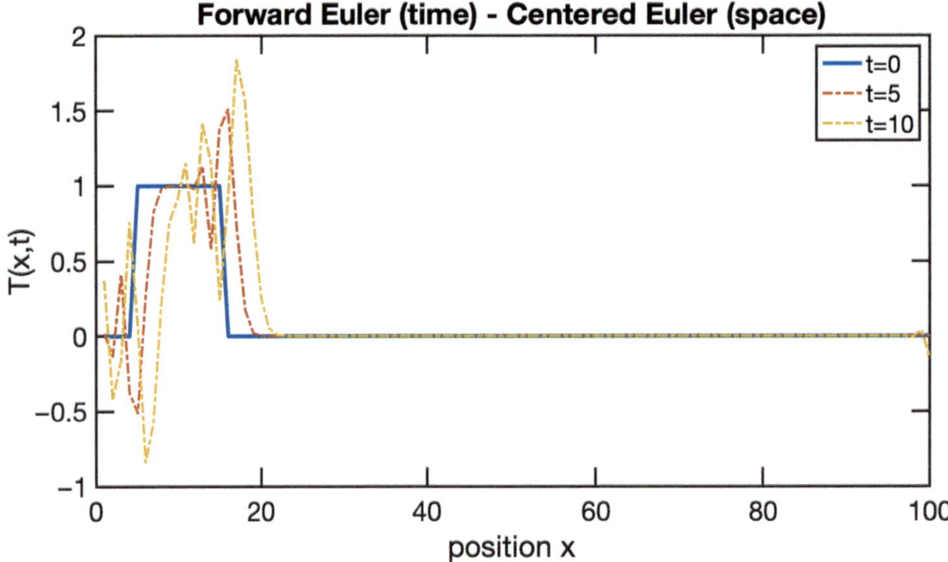

Figure 14.1 Outputs from the first 10 iterations using the forward Euler (time) and centered Euler (space) finite difference scheme for the linear advection equation with the Courant number $C = 0.5$.

After reorganizing the equation we get

$$\frac{T_k^{n+1} - T_k^n}{\Delta t} = \frac{\partial T}{\partial t}|_{t_n} + \underbrace{\frac{\Delta t}{2}\frac{\partial^2 T}{\partial t^2}|_{t_n} + \mathcal{O}(\Delta t^2)}_{truncation\ in\ time\ \mathcal{T}_t}. \tag{14.12}$$

The spatial component of the truncation error $\mathcal{T}_x$ requires two expansions:

$$T_{k+1}^n = T_k^n + \frac{\partial T}{\partial x}|_{x_k}\Delta x + \frac{\Delta x^2}{2}\frac{\partial^2 T}{\partial x^2}|_{x_k} + \frac{\Delta x^3}{6}\frac{\partial^3 T}{\partial x^3}|_{x_k} + \mathcal{O}(\Delta x^4), \tag{14.13}$$

$$T_{k-1}^n = T_k^n - \frac{\partial T}{\partial x}|_{x_k}\Delta x + \frac{\Delta x^2}{2}\frac{\partial^2 T}{\partial x^2}|_{x_k} - \frac{\Delta x^3}{6}\frac{\partial^3 T}{\partial x^3}|_{x_k} + \mathcal{O}(\Delta x^4). \tag{14.14}$$

Substracting these two equations and dividing by $2\Delta x$ yields

$$\frac{T_{k+1}^n - T_{k-1}^n}{2\Delta x} = \frac{\partial T}{\partial x}|_{x_k} + \underbrace{\frac{\Delta x^2}{3}\frac{\partial^3 T}{\partial x^3}|_{x_k} + \mathcal{O}(\Delta x^4)}_{truncation\ in\ space\ \mathcal{T}_x}. \tag{14.15}$$

We can use (Eq. 14.12) and (Eq. 14.15) to rewrite our FDE in the following form

$$\frac{T_k^{n+1} - T_k^n}{\Delta t} + u\frac{T_{k+1}^n - T_{k-1}^n}{2\Delta x} = \frac{\partial T}{\partial t} + u\frac{\partial T}{\partial x} + \underbrace{\frac{\Delta t}{2}\frac{\partial^2 T}{\partial t^2}|_{t_n} + \frac{\Delta x^2}{3}\frac{\partial^3 T}{\partial x^3}|_{x_k} + \mathcal{O}(\Delta x^4, \Delta t^2)}_{\mathcal{T}=\mathcal{T}_t+\mathcal{T}_x}. \tag{14.16}$$

The scheme is consistent as the truncation error vanishes with $\Delta t, \Delta x \to 0$. The scheme is first order in time and second order in space. Interestingly, the leading term in space of the truncation error depends on the third-order derivative of T with respect to the position x. We saw in Chapter 13 that equations of the form

$$\frac{\partial T}{\partial t} = \alpha \frac{\partial^3 T}{\partial x^3} \tag{14.17}$$

are dispersive. The leading term of the spatial component of the truncation error therefore carries numerical dispersion into our numerical solution, which is observed by the ringing (oscillations) observed in the solution at early times (Figure 14.1).

14.1.4 **Stability**

The numerical scheme is unstable, irrespective to the choice of Δt (unconditionally unstable), and we show why here. The argument is based on the analysis of the truncation error derived in Section 14.1.3, more specifically on the leading-order term in time

$$-\frac{\Delta t}{2}\frac{\partial^2 T}{\partial t^2}.$$

We use the definition of the PDE we are aiming to solve to replace the time derivatives with spatial derivatives

$$\frac{\partial T}{\partial t} = -u\frac{\partial T}{\partial x}. \tag{14.18}$$

That allows us to transform the second-order time derivative according to

$$\frac{\partial^2 T}{\partial t^2} = \frac{\partial}{\partial t}\left(-u\frac{\partial T}{\partial x}\right). \tag{14.19}$$

Assuming the velocity u does not vary over time and that the function T is smooth and allows us to exchange the order of the derivatives (twice differentiable and continuous), we can transform this term further

$$\frac{\partial}{\partial t}\left(-u\frac{\partial T}{\partial x}\right) = -u\frac{\partial}{\partial x}\left(\frac{\partial T}{\partial t}\right) = u^2\frac{\partial^2 T}{\partial x^2}. \tag{14.20}$$

Rewriting our FDE (Eq. 14.16) with the substitution for the second derivative over time leads to

$$\frac{T_k^{n+1} - T_k^n}{\Delta t} + u\frac{T_{k+1}^n - T_{k-1}^n}{2\Delta x} = \frac{\partial T}{\partial t} + u\frac{\partial T}{\partial x} + \frac{u^2\Delta t}{2}\frac{\partial^2 T}{\partial x^2}|_{t_n} + \frac{\Delta x^2}{3}\frac{\partial^3 T}{\partial x^3}|_{x_k} + \mathcal{O}(\Delta x^4, \Delta t^2). \tag{14.21}$$

Among all these terms, one stands out as the culprit for the unstable behavior of the numerical solution. It is more obvious if we focus on the second spatial derivative in the truncation error and how it behaves when taken in conjunction with the partial derivative over time of T

$$\frac{\partial T}{\partial t} \propto -\frac{u^2\Delta t}{2}\frac{\partial^2 T}{\partial x^2}, \tag{14.22}$$

which introduces a numerical diffusion term into the FDE. The effective diffusion coefficient for this numerical diffusion term is $D = -u^2\Delta t/2$, which is negative for all $\Delta t > 0$ and introduces

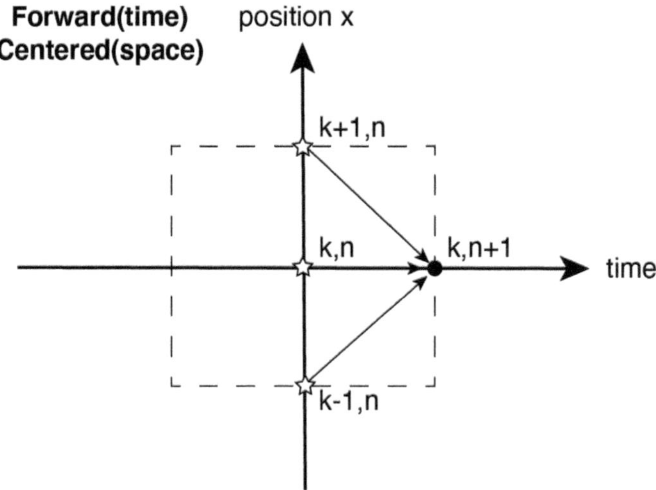

Figure 14.2 Stencil for the forward (time) and centered (space) scheme for the advection equation. The slope of the arrows indicate the characteristic velocity at which information propagates in this finite difference scheme.

negative numerical diffusion, which is always unstable. In other words, the truncation error for the FE–CE is plagued by two issues: numerical dispersion and (more importantly) negative numerical diffusion. The latter makes this scheme unconditionally unstable for the linear advection equation.

14.1.5 Lack of convergence and back to the drawing board

The FE–CE scheme is not stable for any Δt and therefore never converges to the solution of the PDE. A somehow naive qualitative understanding of the failing of the scheme can be gained with a visual representation of the stencil of the FE–CE model for the advection equation.

Figure 14.2 shows that information propagates from the time t_n to t_{n+1} along three directions (symmetric) with arrows connecting positions x_{k-1}, x_k, and x_{k+1} to x_k at the next time step. Each of these arrows can be characterized by their slope in this diagram, which provides a measure for the characteristic velocity at which information propagates in our scheme (i.e., interactions between neighbor nodes). For instance, the arrows originating (at t_n) from x_{k+1}, x_k, and x_{k-1} correspond to velocities $-\Delta x/\Delta t$, 0 and $\Delta x/\Delta t$. In other words, the information is traveling symmetrically with both negative and positive velocities with the FE–CE model. Is this consistent with the idea of the advection equation? Consider an arm-wavy discussion for the sake of the argument. Imagine that the advection equation is a model for how an olfactive signal travels through the air because of wind. The information encoded in that signal (presence of a prey) travels with the wind to a predator. The predator will only know of the presence of a nearby prey (assuming it relies solely on its sense of smell) if it is downwind from it, that is, if the wind (and therefore information) flows from the prey to the predator. Along the same lines, in our scheme above (and the diagram in Figure 14.2), we see that the new point (x_k, t_{n+1}) can only be downwind of one of the two arrays originating from (x_{k+1}, t_n) and (x_{k-1}, t_n). It would be the former if the velocity of the advection equation $u < 0$ and the latter if $u > 0$. That means that the centered finite difference scheme presented here always includes information traveling against the direction of the "wind" or velocity u. Resolving this issue leads to the definition of another finite difference approximation: the **upwind scheme.**

14.2 Scheme 2: Upwind scheme

The main idea behind the upwind scheme is to build from the previous discussion and design a finite difference scheme where the spatial discretization depends on the sign of the velocity u to make sure that the flow of information is consistent with the advection velocity u. The scheme will therefore no longer be symmetric in space (see Figure 14.3) and its implementation requires a knowledge of the sign of u.

14.2.1 Implementation

The FDE for the upwind scheme requires knowledge of the sign of the velocity u

$$\frac{T_k^{n+1} - T_k^n}{\Delta t} = -u \begin{cases} \frac{T_k^n - T_{k-1}^n}{\Delta x} & \text{if} \quad u > 0, \\ \frac{T_{k+1}^n - T_k^n}{\Delta x} & \textit{otherwise.} \end{cases} \tag{14.23}$$

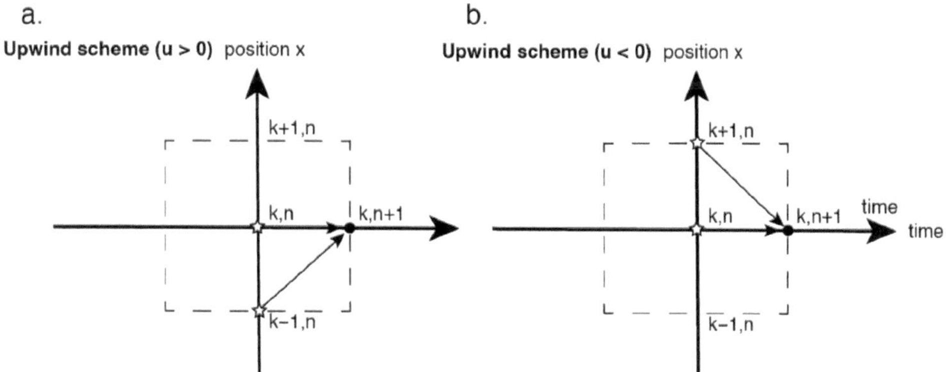

Figure 14.3 Stencil for the upwind scheme for the advection equation. The spatial part of the figure depends on the sign of the velocity u in the advection equation.

The updated pseudo-code reads

```
dt=...;
dx=...;
u=...;
C=u*dt/dx;
T_old=...; % Initial condition
for time=1:maxtime-1 % Marching equation
        for position=2:N-1
                if C>0
                    T(position)=T_old(position)+... \\
         C*(T_old(position)-T_old(position-1));
         else
                   T(position)=T_old(position)+... \\
         C*(T_old(position+1)-T_old(position));
         end
   end
% Boundary conditions to complete T(1) and T(N)
...
end
```

14.2.2 **Performance**

Repeating the same exercise as in Section 14.1.2, again with a set Courant number $C = u\Delta t/\Delta x = 0.5$, we obtain the results shown in Figure 14.4. We readily observe that the problem with numerical dispersion and stability seems to be resolved. However, it is clear that the numerical solution is far from ideal, as the amplitude, width, and overall shape of the rectangular wave are not preserved.

So far we have not tested the results for different choices of C that we can set through the coarseness of the discretization Δt and Δx. We now compare the numerical results in Figure 14.4 to two other cases where $C = 1, 1.2$. Setting the discretization such that $\Delta t = \Delta x/u$ (i.e., $C = 1$) leads to a drastically different outcome (see Figure 14.5).

The results for $C = 1$ are excellent, they preserve the shape and amplitude of the wave as it propagates to the right of the domain. Interestingly, because of the periodic boundary conditions, after 100 iterations, the wave reaches the position it started from and the overlap there shows clearly that the scheme performs very well. It is interesting to reflect on how different the numerical solutions are for $C = 0.5$ and 1. Pushing forward and exploring yet another value of $C = 1.2$, we obtain the results displayed in Figure 14.6. It is clear that the solution here is not stable and it seems that we face numerical dispersion again (i.e., ringing of the numerical solution).

Before addressing the stability conditions that would explain the behavior of the numerical solution, we derive the truncation error and establish the consistency of the upwind scheme for the linear advection equation.

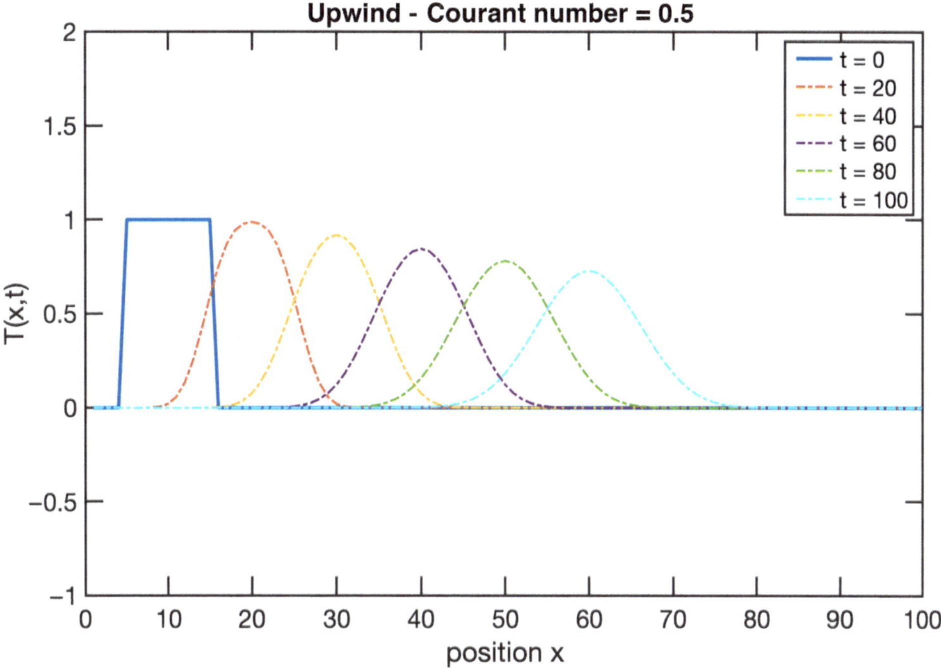

Figure 14.4 Outputs from the first 100 iterations using the upwind finite difference scheme for the linear advection equation with the Courant number $C = 0.5$. Note that the numerical solution is stable although it does not preserve the shape and amplitude of the initial rectangular wave.

14.2.3 **Consistency**

To better contrast the upwind scheme with the FE–CE scheme, we first recast the FDE for the upwind scheme (assuming $u > 0$) as

$$\frac{T_k^{n+1} - T_k^n}{\Delta t} + u\frac{T_k^n - T_{k-1}^n}{\Delta x} = \underbrace{\frac{T_k^{n+1} - T_k^n}{\Delta t} + u\frac{T_{k+1}^n - T_{k-1}^n}{2\Delta x}}_{Forward(time)-Centered(space)} - u\frac{T_{k+1}^n - 2T_k^n + T_{k-1}^n}{2\Delta x}. \tag{14.24}$$

We have already derived the truncation error associated with the FE–CE scheme (Eq. 14.21), so we just need to focus on the last term that is implicitly added by the upwind discretization

$$\Lambda \equiv -u\frac{T_{k+1}^n - 2T_k^n + T_{k-1}^n}{2\Delta x}. \tag{14.25}$$

Using two Taylor expansions for $T(x,t)$ around x_k and t_n we get

$$T_{k+1}^n = T_k^n + \Delta x\frac{\partial T}{\partial x}|_{x_k} + \frac{\Delta x^2}{2}\frac{\partial^2 T}{\partial x^2} + \frac{\Delta x^3}{6}\frac{\partial^3 T}{\partial x^3}|_{x_k} + \mathcal{O}(\Delta x^4) \tag{14.26}$$

$$T_{k-1}^n = T_k^n - \Delta x\frac{\partial T}{\partial x}|_{x_k} + \frac{\Delta x^2}{2}\frac{\partial^2 T}{\partial x^2} - \frac{\Delta x^3}{6}\frac{\partial^3 T}{\partial x^3}|_{x_k} + \mathcal{O}(\Delta x^4). \tag{14.27}$$

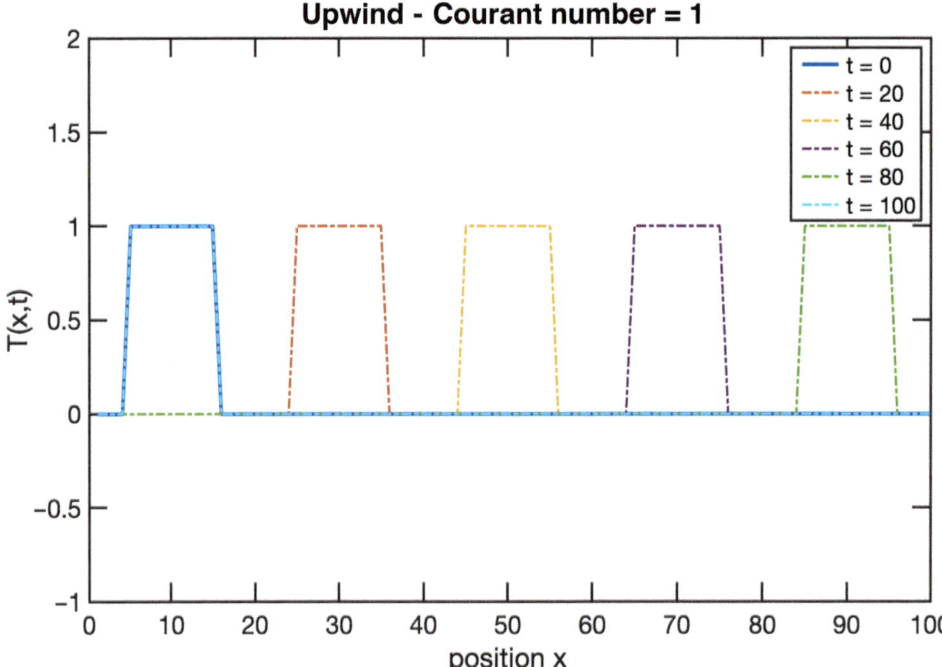

Figure 14.5 Outputs from the first 100 iterations using the upwind finite difference scheme for the linear advection equation with the Courant number $C = 1$. Note here that because of the periodic boundaries, the wave reaches its initial position after 100 iterations. It is clear here that the shape and amplitude of the wave is preserved throughout the simulation.

Adding these two equations and reorganizing the terms leads to

$$\frac{T_{k+1}^n - 2T_k^n + T_{k-1}^n}{2\Delta x} = \frac{\Delta x}{2}\frac{\partial^2 T}{\partial x^2} + \mathcal{O}(\Delta x^3), \tag{14.28}$$

which we can now add to the truncation error for the FE–CE scheme to obtain the overall truncation error for the upwind scheme

$$\frac{T_k^{n+1} - T_k^n}{\Delta t} + u\frac{T_k^n - T_{k-1}^n}{\Delta x} = \frac{\partial T}{\partial t} + u\frac{\partial T}{\partial x} + \frac{u^2\Delta t - \mathbf{u\Delta x}}{2}\frac{\partial^2 T}{\partial x^2}\Big|_{t_n} + \frac{\Delta x^2}{3}\frac{\partial^3 T}{\partial x^3}\Big|_{x_k} + \mathcal{O}(\Delta x^3, \Delta t^2). \tag{14.29}$$

The new term proper to the upwind scheme is in bold type in equation 14.29. We consider the truncation error in this equation in more detail in Section 14.2.4. We note however that because of the new term the upwind scheme is now first order in space, and it is still first order in time. The upwind scheme is consistent because the truncation error vanishes as $\Delta t, \Delta x \to 0$.

14.2.4 **Stability**

We can leverage the work done when studying the stability of the FE–CE scheme to establish the stability condition for the upwind scheme. For the former, we saw that a negative diffusion term

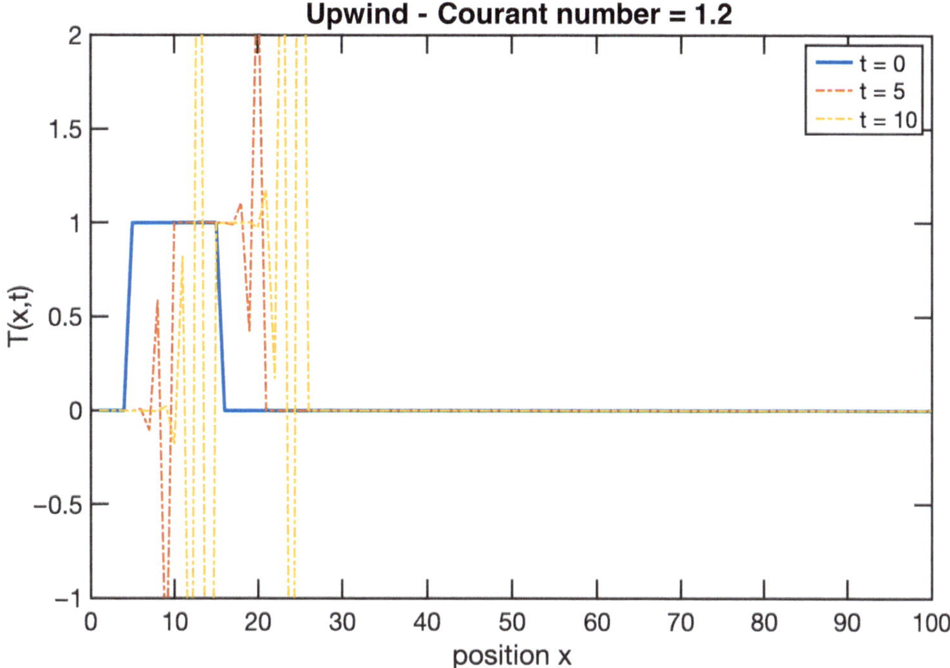

Figure 14.6 Outputs from the first 10 iterations using the upwind finite difference scheme for the linear advection equation with the Courant number $C = 1.2$. We observe a similar outcome as for the forward Euler (time)–centered Euler (space) scheme.

in the truncation error was responsible for the scheme to be unstable irrespective to the choice of $\Delta x, \Delta t$. From (Eq. 14.29) we see that the numerical diffusion term is now

$$\frac{\partial T}{\partial t} \propto -\frac{u^2\Delta t - u\Delta x}{2}\frac{\partial^2 T}{\partial x^2}, \tag{14.30}$$

so that the numerical diffusion coefficient is defined by

$$D = \frac{1}{2}\left(u\Delta x - u^2\Delta t\right) = \frac{u}{2}\left(\Delta x - u\Delta t\right), \tag{14.31}$$

which is positive (therefore stable) when $C \leq 1$ and negative (unstable) if $C > 1$. Interestingly, we note that the numerical diffusion term cancels if $C = 1$, which explains why the wave shape and amplitude were preserved in that particular case. We can relate now the sign of the numerical diffusion term to the stability observed with the upwind scheme and the special case when $C = 1$.

14.3 Application to cosmogenic nuclides formation during sedimentation or erosion

A very nice application of the linear advection equation is the determination of the long-term (scale of tens to hundreds of thousands of years at least) average erosion or sedimentation rates

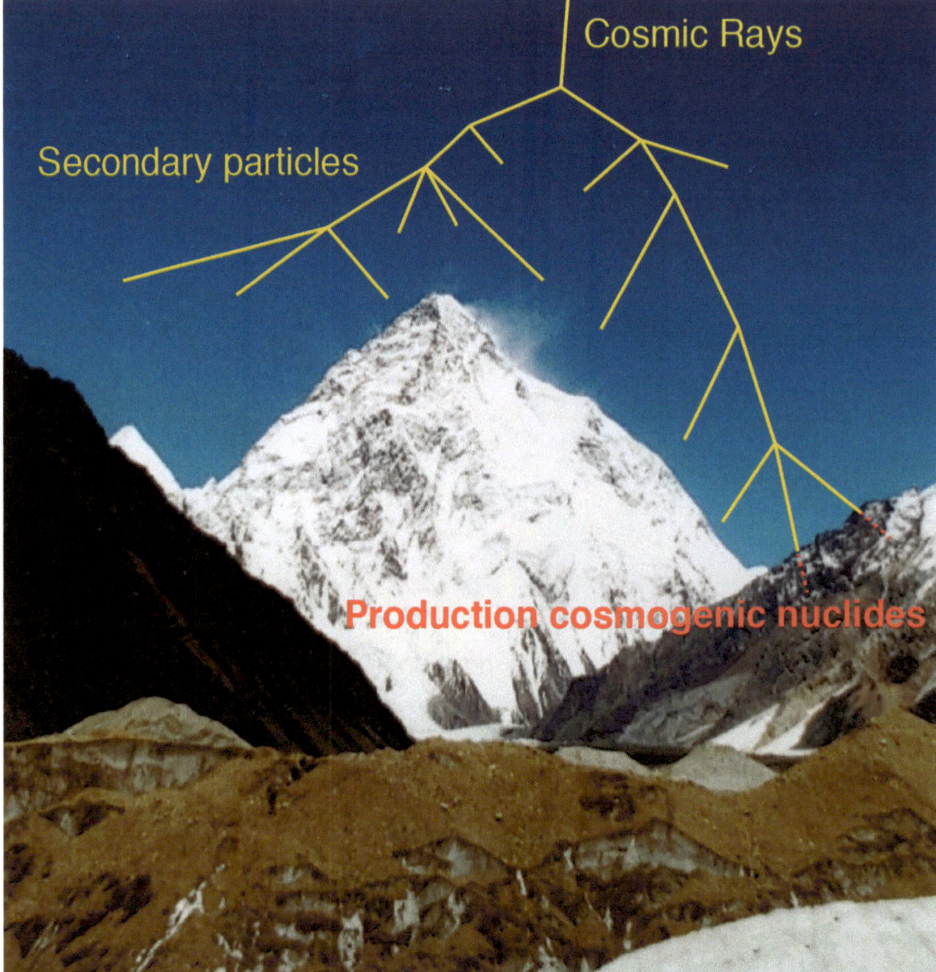

Figure 14.7 Schematic illustration of cosmic ray interaction with the atmosphere to produce secondary particles and produce a cosmic shower that leads to the production of cosmogenic nuclides at the Earth's surface. For those interested, the mountain in the background is K2 and the picture was taken by the author in July 2001.

from the distribution of cosmogenic nuclides in soils. For more information about the underlying processes and methods used in this important subfield of geomorphology, interested readers are directed to [2]. We provide here a concise introduction for both the processes upon which this method is based and the derivation of the equation. We then derive an analytical expression for steady-state profiles to test the performance of our upwind FDE.

Cosmic rays are continuously produced by the Sun and supernovae outside the solar system. These rays consist of electrically charged particles (e.g., protons, electrons, and leptons). As these cosmic rays interact with the upper atmosphere, collisions/reactions with atoms in the atmosphere lead to the formation of secondary particles that form what is often called a cosmic shower (see Figure 14.7). A substantial fraction of the secondary particles produced that reach the Earth's surface are muons, which can react with rocks at and near the surface. These muon particles interact with minerals and generate reactions where radioactive atoms are produced. As muons exchange energy with the soils they penetrate, they can only generate these reactions within a certain depth range (skin depth) before exhausting their energy. The skin depth depends on the energy of the muon and the density of the soil. Interestingly, because the Earth's magnetic field is not homogeneous (i.e., the dominant dipole component creates strong latitudinal variations) and because secondary particles lose energy as they interact with atoms in the atmosphere, the production rate of cosmogenic nuclides J (these radioactive atoms produced in the soil) depends not only on the mineralogy of the soil, but also on the latitude and elevation of the site.

In summary, as the muons penetrate the soil they lose energy and their ability to produce cosmogenic nuclides decreases exponentially with depth z

$$J(z) = J_0 \exp(-\mu z), \tag{14.32}$$

where J_0 is a surface production rate (i.e., depends on altitude and latitude of the site) and $1/\mu$ is a characteristic penetration depth that depends on the soil and the average energy of the incoming muons.

The cosmogenic nuclides generated by this process are radioactive and assuming a static (i.e., no sedimentation nor erosion of the soil) system, the concentration of nuclides at time t and position z follows

$$\frac{dC}{dt} = -\lambda C + J(z), \tag{14.33}$$

with λ the decay constant for the radioactive nuclide of interest (for example, ^{10}Be). This equation admits a simple steady-state solution

$$C(z) = \frac{J_0}{\lambda} \exp(-\mu z). \tag{14.34}$$

If we now relax the assumption that the soil experiences no net erosion or sedimentation then we can expand the time derivative of C as in (Eq. 14.6) to get

$$\frac{dC}{dt} = \frac{\partial C}{\partial t} + u\frac{\partial C}{\partial z} = -\lambda C + J_0 \exp(-\mu z), \tag{14.35}$$

which is the mass conservation equation that governs the evolution of the mass (or number of atoms per unit mass of soil) of cosmogenic nuclides produced in the soil. In this equation, the velocity u parameterizes the movement of a reference volume of soil with respect to the surface boundary (see Figures 14.8 and 14.9). As such, because the sign of the velocity u changes if we consider erosion versus sedimentation, we must pay special attention to the upwind discretization of the PDE.

For sedimentation ($u > 0$) the upwind FDE becomes

$$\frac{C_k^{n+1} - C_k^n}{\Delta t} + u\frac{C_k^n - C_{k-1}^n}{\Delta z} = -\lambda C_k^n + J_0 \exp(-\mu k \Delta z), \tag{14.36}$$

with a boundary condition that comes into play when calculating C_1^n

$$\frac{C_1^{n+1} - C_1^n}{\Delta t} + u\frac{C_1^n - C_s}{\Delta z} = -\lambda C_1^n + J_0 \exp(-\mu \Delta z), \tag{14.37}$$

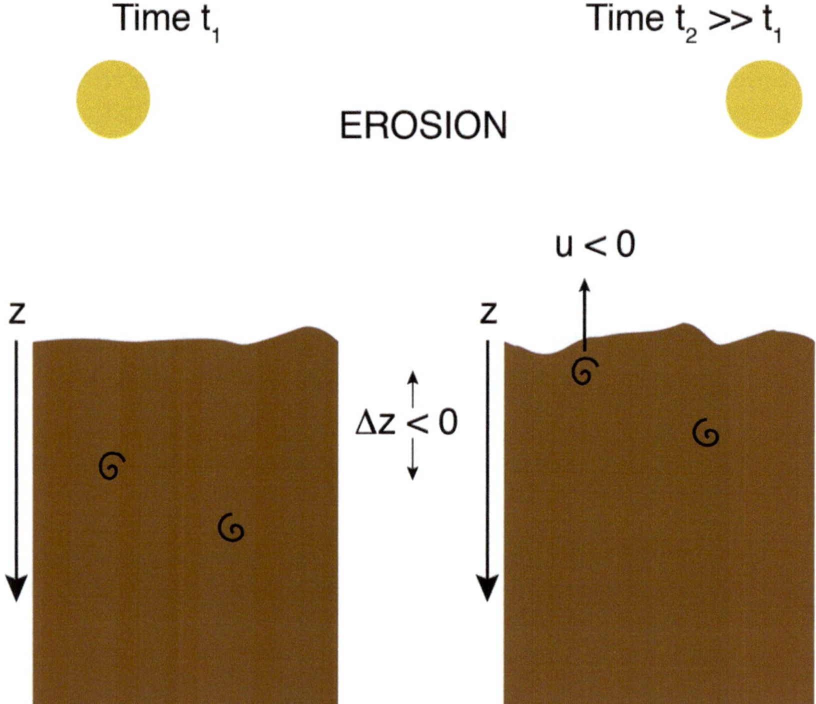

Figure 14.8 Schematic illustration for the erosion model, which relates the velocity $u < 0$ to the erosion rate.

and C_s is the surface concentration, which if the sedimentation rate is fast enough can be approximated as $C_s = 0$, that is, freshly deposited sediment with no previous production of cosmogenic nuclides. We also must specify an initial condition $C(z, t = 0) = C_0(z)$ to have a complete set of equations to solve this problem. In the following solutions, $C_0 = 0$, which is not necessarily the most realistic initial condition, we could have selected for example the steady-state profile solution for the static soil column (Eq. 14.34). The analytical solution for the concentration profile with sedimentation ($u > 0$) under steady-state is[1]

$$C(z) = C_0 \exp\left(-\frac{\lambda z}{u}\right) + \frac{J_0}{\lambda - \mu u}\left[\exp(-\mu z) - \exp\left(-\frac{\lambda z}{u}\right)\right]. \tag{14.38}$$

We can now compare the numerical solution with the upwind scheme defined in (Eq. 14.36) to this analytical expression by running the code over a time scale that is large enough (significantly greater than the half-life of the nuclide, here about 1.5 Myrs for ^{10}Be), to reach a steady-state (here we selected 50 Myrs). The production rate at the surface is here set to $J_0 = 6$ atom/g/yr and the skin depth $1/\mu = 63$ in cm, lastly the sedimentation rate is $u = 10^{-4}$ cm/yr. The results are shown in Figure 14.10.

[1] This analytical expression for $0 = u\frac{\partial C}{\partial z} + \lambda C - J_0 \exp(-\mu z)$ can be obtained by the method of the variation of the constant. See the example in Part I Chapter 5.

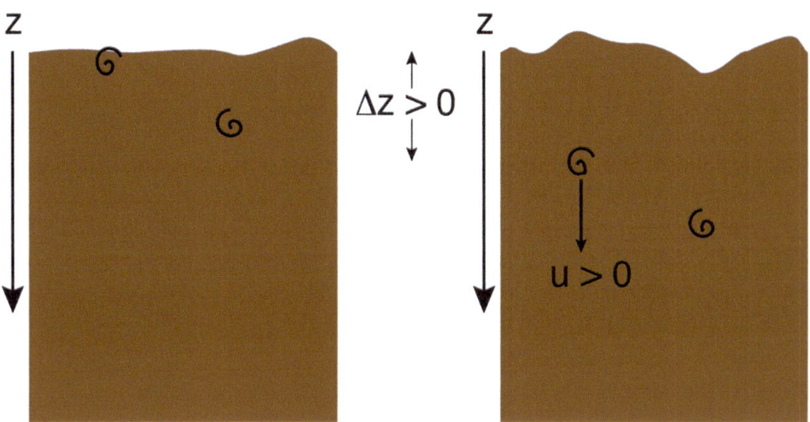

Figure 14.9 Schematic illustration for the sedimentation model, which relates the velocity $u > 0$ to the sedimentation rate.

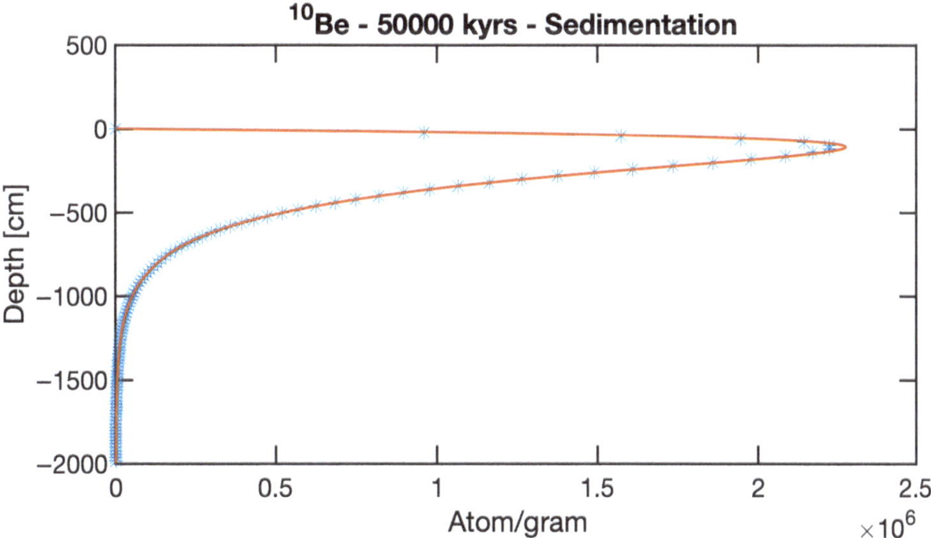

Figure 14.10 Profile of ^{10}Be concentration profile in soils subjected to a fixed sedimentation rate. The numerical solution (stars) is compared to the analytical solution (solid line). The sedimentation rate is constant ($u = 10^{-4}$ cm/yr) and we use a Courant number $C = 0.25$.

Considering now erosion ($u < 0$), we need a different FDE with our upwind scheme

$$\frac{C_k^{n+1} - C_k^n}{\Delta t} + u\frac{C_{k+1}^n - C_k^n}{\Delta z} = -\lambda C_k^n + J_0 \exp(-\mu k \Delta z), \tag{14.39}$$

and the boundary condition is now $C(z \gg 1/\mu) = 0$. We use the same initial condition for the test of our numerical scheme ($C_0(z) = 0$). Considering the steady-state nuclide concentration profile, there is an analytical solution given by

$$C(z) = \frac{J_0 \exp(-\mu z)}{\mu|u| + \lambda}. \tag{14.40}$$

Again the numerical solution and the analytical expression using the same parameters as before but with $u = -10^{-4}$ cm/yr are in excellent agreement. Note that a code that would be able to deal with either sedimentation or erosion case or, for example, conditions that over time transition from net sedimentation to erosion, or vice versa, would require a special treatment and conditional tests to adapt the FDE to the sign of the velocity u. This is not ideal and is one motivation for deriving a (conditionally) stable finite difference approach that is symmetric in space and therefore does not need special treatment if the velocity field alters. We discuss the symmetric (Lax–Wendroff) approach in Section 14.4.

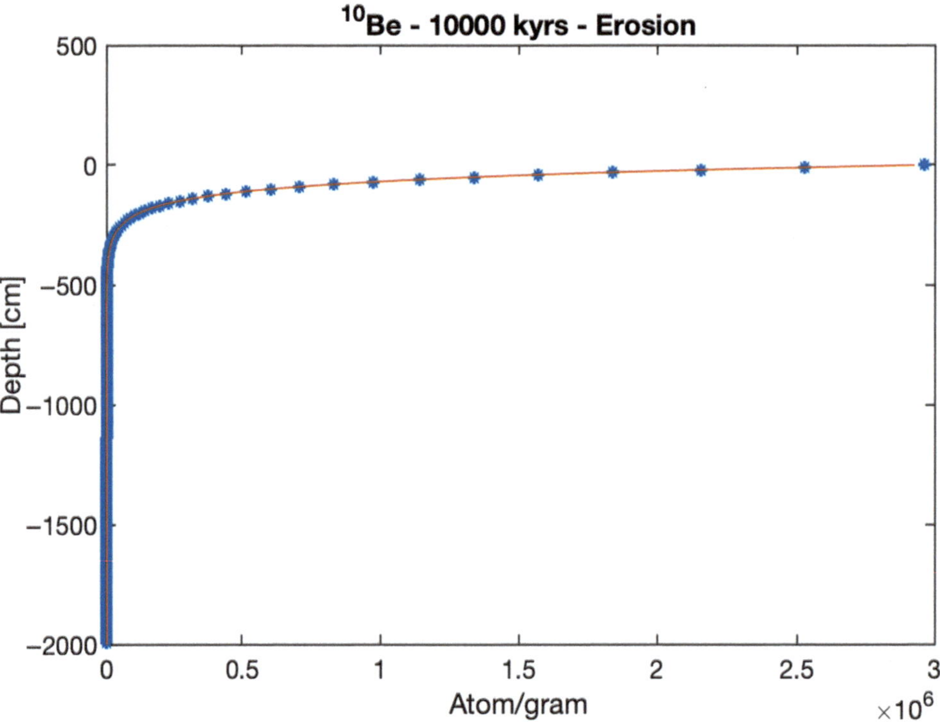

Figure 14.11 Profile of ^{10}Be concentration profile in soils. The results from the upwind scheme (markers) are compared to the analytical solution (solid line) for a constant erosion rate ($u = -10^{-4}$ cm/yr) using a Courant number (absolute value) $C = 0.25$.

14.4 **Lax–Wendroff method**

Among the multiple versions of Lax–Wendroff methods, we focus here on the two-step Richtmyer approach. It is a two-step method because the integration over time involves two half-steps (duration $\Delta t/2$). It is built as a conservative method for hyperbolic PDEs and is second-order accurate in space and time. In finite differences, unlike finite volume methods, local conservation is not guaranteed (see Part IV), that is, truncation errors can lead to a mass or energy balance that can drift away from the target (not a closed budget). However, there are ways to develop conservative methods with finite differences, generally based on a proper discretization for the fluxes $q(T)$ in

$$\frac{\partial T(x,t)}{\partial t} + \frac{\partial (qT(x,t))}{\partial x} = 0. \tag{14.41}$$

The interested reader is encouraged to read [17] for more information about conservative approaches with finite difference numerical methods.

The two-step Richtmyer approach considers a first step over a time increment $\Delta t/2$ and estimates of the field of interest (here T) at the half points between position $x_k = k\Delta x$ and its nearest neighbors

$$T^{n+1/2}_{k+1/2} = \frac{1}{2}\left[T^n_{k+1} - T^n_k\right] - \frac{\Delta t}{2\Delta x}\left[q(T^n_{k+1} - q(T^n_k)\right] \tag{14.42}$$

$$T^{n+1/2}_{k-1/2} = \frac{1}{2}\left[T^n_k - T^n_{k-1}\right] - \frac{\Delta t}{2\Delta x}\left[q(T^n_k - q(T^n_{k-1})\right]. \tag{14.43}$$

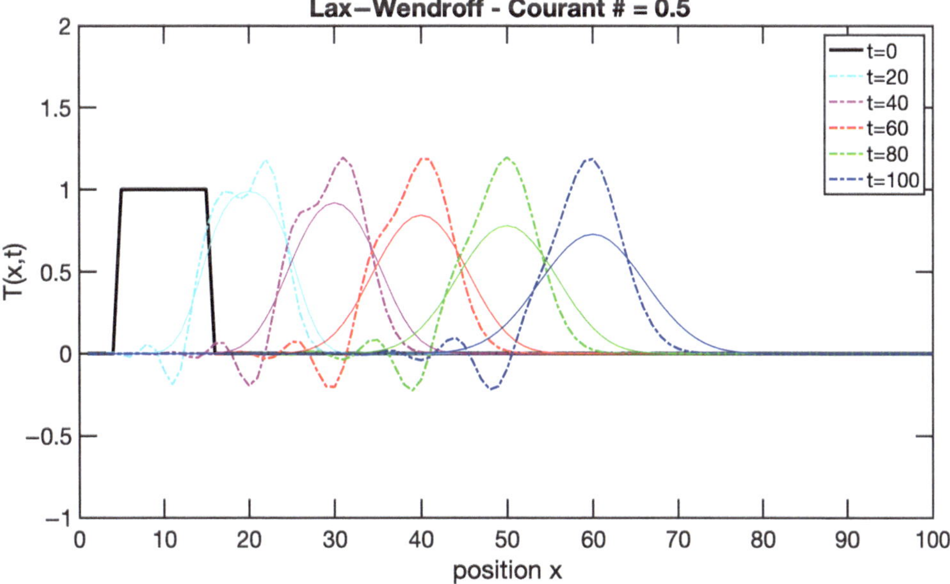

Figure 14.12 Outputs from the first 100 iterations comparing the Lax–Wendroff method to upwind finite difference for the linear advection equation with the Courant number $C = 0.5$. We see that the Lax–Wendroff method (dashed lines) is subject to numerical dispersion (ringing of the solution) and does not preserve shape, but the clear damping of the solution of the upwind scheme (thin solid lines) is no longer an issue. Both methods provide excellent solutions for $C = 1$.

The second step finishes the time integration

$$T_k^{n+1} = T_k^n - \frac{\Delta t}{\Delta x}\left[q(T_{k+1/2}^{n+1/2}) - q(T_{k-1/2}^{n+1/2})\right]. \tag{14.44}$$

For the simplest form of the advection equation (see earlier in this section), the flux is defined as $q = uT(x, t)$ with u taken as a homogeneous ($u \neq u(x)$) velocity across the domain. Although we do not derive the expression for the truncation error of this scheme, it is second order in time and space and it includes a term $\partial^3 T/\partial x^3$ that carries numerical dispersion. The existence of numerical dispersion is obvious from the numerical solution (Figure 14.12). The Lax–Wendroff scheme obeys a similar stability condition as the upwind scheme here, namely that the Courant number $C = u\Delta t/\Delta x \leq 1$, but the scheme is symmetrical in space and therefore can be used without modification for both $u > 0$ and $u < 0$. A major issue with an upwind scheme for the linear advection equation is the presence of numerical diffusion, which damps significantly sharp variations in space of the function $T(x, t)$, the Lax–Wendroff scheme is much less susceptible to this particular issue (see Figure 14.12).

14.5 **Method of characteristics**

The method of characteristics applies to hyperbolic equations such as the linear advection equation presented here. For more details about the method of characteristics and applications, the interested reader is directed to Zachmanoglou and Dale [29]. The idea behind this approach is to transform the PDE (variables are space and time) into a set of parametric ordinary differential equations (ODEs). We show here how to apply it to our example and then discuss methods for numerical solution (based on our work with ODEs). Although the solution procedure is rather simple, the mathematical concepts underlying this approach are a bit more complicated, and it is discussed here because it is a widely used approach and therefore deserves some attention.

Starting with a generalized first-order PDE

$$a(x, t)\frac{\partial T}{\partial t} + b(x, t)\frac{\partial T}{\partial x} = c(x, t), \tag{14.45}$$

with coefficients a, b, and c all possibly depending on time and position. Let us define a generalized gradient operator $\nabla = (\partial/\partial t, \partial/\partial x, -1)^T$, which is a vector of partial derivative with respect to our two primary variables (x, t). We define another vector as a collection of the scalar fields a, b, c with $\mathbf{A} = (a(x, t), b(x, t), c(x, t))^T$ (where ^{T} denotes the transpose of the vector).

If we take a surface as the collection of points $S = t, x, z = T(x, t)$, which describes possible solutions at any time and position for $T(x, t)$, and let $F = z - T(x, t) = 0$ be a parameterization of the surface. Then a normal path to that surface at (x, t) is given by ∇F

$$\nabla F = \begin{pmatrix} -\frac{\partial T}{\partial t} \\ -\frac{\partial T}{\partial x} \\ 1 \end{pmatrix}. \tag{14.46}$$

We can now express (Eq. 14.45) as a dot (scalar) product between two vectors

$$\mathbf{A} \cdot \nabla F = 0. \tag{14.47}$$

Mathematically that means that our first-order PDE (Eq. 14.45) describes a solution for $T(x, t)$ such that the vector $\mathbf{A}$ is normal to ∇F. In other words, $\mathbf{A}$ is tangential to the plane described by $F(x, t) = z - T(x, t) = 0$. So any solution of the PDE should define a plane F such that the vector of parameters $\mathbf{A}$ is tangential to it at any point (x, t). The next step is to parameterize a curve on the surface F, where two new parameters (s, r) are introduced. The parameter s describes how much we travel along the curve from an initial condition (x_0, t_0) and the parameter r describes the choice of initial condition (when $s = 0$), and it could, at this stage, be any choice (x_0, t_0). We can define characteristic curves that describe tangent to the curve at any point given by the parameters (s, r) with the following ODEs

$$\frac{dt}{ds} = a(x(s, r), t(s, r)) \tag{14.48}$$

$$\frac{dx}{ds} = b(x(s, r), t(s, r)) \tag{14.49}$$

$$\frac{dz}{ds} = c(x(s, r), t(s, r)). \tag{14.50}$$

Each of these equations can be integrated either numerically (using any method we used for ODEs in Part II of the book) or analytically if the expressions for $a(x, t)$, $b(x, t)$, and $c(x, t)$ are simple enough. For the sake of developing a complete and simple work flow through this problem we assume that the parameters a, b, and c are constant (we will set $a = 1$, $b = u$, and $c = 0$ to retrieve our linear advection equation) and integration leads to

$$t(s, r) = s + k_1(r) \tag{14.51}$$

$$x(s, r) = us + k_2(r) \tag{14.52}$$

$$z(s, r) = k_3(r). \tag{14.53}$$

Using the first two equations here, we can isolate the parameters s, r and get their dependence on x, t

$$s = t \tag{14.54}$$

$$r = x - ut. \tag{14.55}$$

Replacing now the expression for $r = x - ut$ into $z(s, r)$ we get an expression for the general solution to the linear advection problem $z(x, t) = T(x, t)$

$$z(x, t) = T(x, t) = k_3(r) = k_3(x - ut), \tag{14.56}$$

where $k_3(x)$ is a function that depends on the choice of initial condition (it represents the collection of points that define the initial condition, i.e., our rectangular wave), which is readily apparent if we take $t = 0$

$$z(x, 0) = T(x, 0) = k_3(x). \tag{14.57}$$

In essence, by some properties of vector calculus, the first-order PDE is transformed into a collection of parameterized ODEs that define a characteristic curve, and each can be solved using methods proper to ordinary equations. This parameterized curve is arbitrary but it exists on the surface that describes the solution to the PDE and therefore allows us to solve for it in a rather elegant way.

14.5.1 Exercises

1. Using parameters listed in Section 14.3, implement an upwind model to solve for the erosion scenario (fixed rate) discussed. Compare your results with those plotted above (you may need to use a high spatial resolution with Δz of a few centimeters).
2. Using the same code, modify it so that it corrects automatically for the sign of the velocity u and allows you to solve for sedimentation or erosion. Be careful with the treatment of the boundary condition.
3. Write a new Matlab script to solve the same problem with the Lax–Wendroff method. Observe that you no longer have to pay special attention to the sign of the velocity u.

Chapter 15

Diffusion equation

A simple one-dimensional derivation of the diffusion equation is presented earlier in Section III. The diffusion equation is ubiquitous in Earth and planetary sciences because it idealizes how a process near to but not at equilibrium relaxes toward an equilibrium or steady-state (which are not the same!). As such, it is commonly used in a broad range of applications, including:

- diffusion chronometry in minerals
- porous media flows (groundwater equation in a confined aquifer)
- heat transfer in the lithosphere
- hillslope evolution models (sediment transport on shallow slopes)
- kinetically limited geochemical transport models.

15.1 The partial differential equation (PDE)

Our target here is to solve numerically the equation

$$\frac{\partial T}{\partial t} = D\frac{\partial^2 T}{\partial x^2}, \tag{15.1}$$

which assumes a homogeneous transport coefficient D. We also discuss some important modifications to our numerical solution procedure when considering a nonhomogeneous D field at the end of this section. Lastly, Eq. 15.1 is not a complete statement of the problem; it needs to be completed with an initial condition (IC) and two boundary conditions (BCs), generally specifying either the field T or gradient/flux at the boundary(ies). BCs on the T are often referred to as Dirichlet boundaries, while flux boundaries are referred to as Neumann BCs. We look at both below.

15.2 Scheme 1: Explicit (FE/time) approach

The first scheme we investigate is perhaps the simplest. It is called an explicit scheme because each of the algebraic equations resulting from the discretization of the PDE involves only a single unknown (it can therefore be solved as individual marching equation solving sequentially for each position). This scheme is shown in Figure 15.1 and the finite difference equation (FDE) is

$$\frac{T_k^{n+1} - T_k^n}{\Delta t} = D\frac{T_{k+1}^n - 2T_k^n + T_{k-1}^n}{\Delta x^2}. \tag{15.2}$$

Introduction to Numerical Modeling in Earth and Planetary Sciences, Christian Huber, Oxford University Press.
© Christian Huber (2025). DOI: 10.1093/oso/9780198802716.003.0016

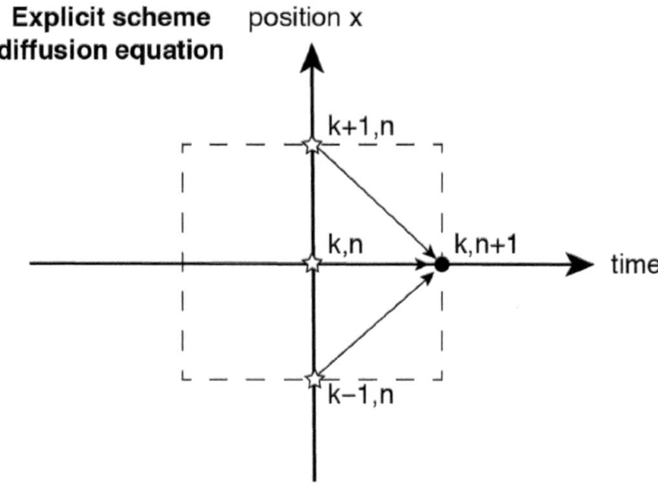

Figure 15.1 The explicit (forward in time) scheme for the diffusion equation.

15.2.1 **Implementation**

The implementation of the explicit (forward Euler, FE) scheme is rather straight-forward. The FDE for each position k can be reframed as (using again superscript for time and subscript for position)

$$T_k^{n+1} = T_k^n + \underbrace{\frac{D\Delta t}{\Delta x^2}}_{Diffusion\ \#\ \mathcal{D}} \left[T_{k+1}^n - 2T_k^n + T_{k-1}^n\right], \quad k = 1, \ldots, \tag{15.3}$$

where we defined the diffusion number $\mathcal{D} = D\Delta t/\Delta x^2$. We consider that the physical domain we want to model spans $x_1 \rightarrow x_N$, and as such we assume x_0 and x_{N+1} to be outside the domain and therefore part of the BCs. Applying (Eq. 15.3) for the first node $k = 1$ yields

$$T_1^{n+1} = T_1^n + \frac{D\Delta t}{\Delta x^2}\left[T_2^n - 2T_1^n + T_0^n\right]. \tag{15.4}$$

At this stage the value of T_0^n is not prescribed, and we discuss here two types of BCs. First, if we assume that the temperature at x_0 is prescribed $T_0^n = T_0(n\Delta t)$, a Dirichlet BC, then the equation above can be computed to update $T_1^n \rightarrow T_1^{n+1}$. On the other end, if the BC at x_0 is a prescribed flux q_0, then it is possible to use a first-order approximation for q_0 such that

$$q_0^n = -k\frac{T_1^n - T_0^n}{\Delta x} + \mathcal{O}(\Delta x^2). \tag{15.5}$$

where k is a transport coefficient relating flux and diffusion (e.g. thermal conductivity for heat transfer). This allows us to substitute

$$-T_1^n + T_0^n = \frac{q_0^n \Delta x}{k}, \tag{15.6}$$

in equation 15.4, which becomes

$$T_1^{n+1} = T_1^n + \frac{D\Delta t}{\Delta x^2}\left[T_2^n - T_1^n + \frac{q_0^n \Delta x}{k}\right]. \tag{15.7}$$

Similarly for $k = N$, the FDE introduces the other BC

$$T_N^{n+1} = T_N^n + \frac{D\Delta t}{\Delta x^2}\left[T_{N+1}^n - 2T_N^n + T_{N-1}^n\right], \tag{15.8}$$

with either T_{N+1} prescribed (Dirichlet) or, if the heat flux is prescribed instead (Neumann)

$$T_N^{n+1} = T_N^n + \frac{D\Delta t}{\Delta x^2}\left[-\frac{q_N^n \Delta x}{k} - T_N^n + T_{N-1}^n\right], \tag{15.9}$$

with

$$q_N^n = -k\frac{T_{N+1}^n - T_N^n}{\Delta x} + \mathcal{O}(\Delta x^2). \tag{15.10}$$

Again, the FE (time) scheme for the one-dimensional diffusion equation is generally referred to as the explicit scheme, mostly because we can solve independently and sequentially each of the FDEs for each $k = 1, \ldots, N$. A pseudo-code is provided below. The explicit nature is to be contrasted with the BE (time) scheme where each FDE involves more than one unknown and the solution procedure resorts to inverting a matrix.

```
dt=...;
dx=...;
D=...;
Diff_n=D*dt/dx^2;
T_old=...; % Initial condition
for time=1:maxtime-1 % Marching equation
        for position=2:N-1
        T(position)=T_old(position)+... \\
        Diff_n*(T_old(position+1)-2*T_old(position)
        +T_old(position-1));
         end
   end
% Boundary conditions to complete T(1) and T(N)
...
end
```

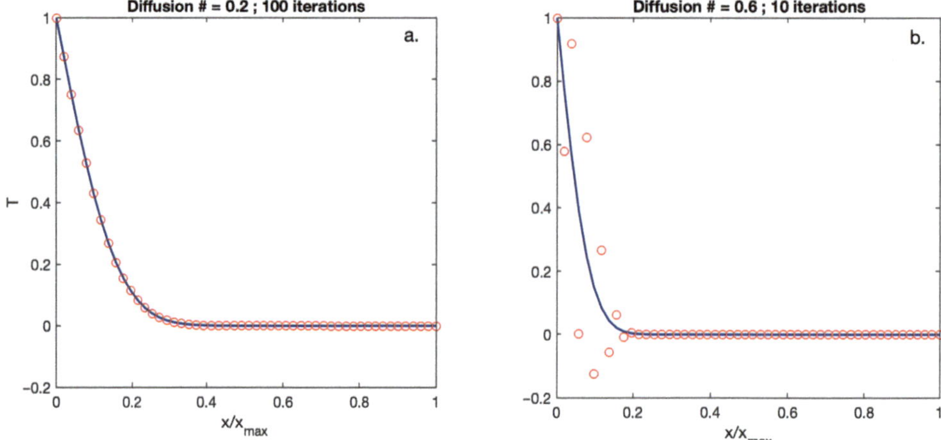

Figure 15.2 Two realizations of the explicit scheme for diffusion. Left panel: The diffusion number $\mathcal{D} = 0.2$ and the simulation is run for 100 iterations. Right panel: The diffusion number is increased to 0.6 and the simulation covers only 10 iterations, as it diverges soon after.

15.2.2 **Performance**

(Eq. 15.3) is the FDE for the explicit scheme and using the definition of the diffusion number $\mathcal{D}$ it can be recast as

$$T_k^{n+1} = T_k^n + \mathcal{D}\left[T_{k+1}^n - 2T_k^n + T_{k-1}^n\right], \quad k = 1, \ldots. \tag{15.11}$$

We can already predict that the performance of the algorithm will depend on the coefficient $\mathcal{D}$ as it encapsulates the information regarding the selected time step and grid spacing. We test the scheme with a simple setup, assuming that the domain $0 \le x \le x_{max}$ is initially at $T(x, 0) = 0$ and the boundary temperature at $x = 0$ is raised instantaneously to $T(0, t) = 1$ ("left" BC). We further assume another Dirichlet BC at x_{max} that $T(x_{max}, t) = 0$. This problem admits an analytical solution in terms of the Fourier series [7]. Moreover, for simulations where the propagation of the heat from the top BC (at $x = 0$) does not reach the other boundary (at x_{max}), the analytical solution reduces to the solution for diffusion in an infinite half space

$$T(x, t) = \operatorname{erfc}\left(\frac{x}{2\sqrt{Dt}}\right), \tag{15.12}$$

where *erfc* is the complementary error function. Two realizations with the explicit model are shown in Figure 15.2, where the left panel shows a simulation with diffusion number $\mathcal{D} = 0.2$ and the right panel shows a shorter run with $\mathcal{D} = 0.6$, and they are compared to the analytical solution. There is a striking difference; the case where $\mathcal{D} = 0.2$ performs very well (very close agreement) while spurious oscillations (even negative temperature values) are observed for $\mathcal{D} = 0.6$. Note that the numerical solution diverged soon after 10 iterations for the latter case. This suggests that the stability of the explicit scheme for diffusion depends directly on the value of $\mathcal{D}$. The next section analyze the convergence of the scheme, starting with the consistency of the explicit scheme.

15.2.3 **Consistency**

Consistency is established here by showing that the limit of the truncation error in time and space goes to 0 when the grid spacing Δx and the time step Δt both $\rightarrow 0$. In other words, consistency is established if the FDE converges to the target PDE (here one-dimensional diffusion) when $\Delta x, \Delta t \rightarrow 0$. Using a Taylor expansion in time around T_k^n to estimate T_k^{n+1} we get

$$T_k^{n+1} = T_k^n + \frac{\partial T}{\partial t}|_n \Delta t + \frac{\partial^2 T}{\partial t^2}\frac{\Delta t^2}{2} + \mathcal{O}(\Delta t^3). \tag{15.13}$$

After reorganizing this equation to retrieve our finite difference approximation in time, we get

$$\frac{T_k^{n+1} - T_k^n}{\Delta t} = \frac{\partial T}{\partial t} + \underbrace{\frac{\Delta t}{2}\frac{\partial^2 T}{\partial t^2} + \mathcal{O}(\Delta t^2)}_{\textit{truncation in time } \mathcal{T}_t}. \tag{15.14}$$

We can now proceed to estimate the truncation error in space similarly to what we did for the upwind scheme for the advection equation, that is, we proceed with two Taylor expansions in space

$$T_{k+1}^n = T_k^n + \frac{\partial T}{\partial x}|_{x_k}\Delta x + \frac{\Delta x^2}{2}\frac{\partial^2 T}{\partial x^2}|_{x_k} + \frac{\Delta x^3}{6}\frac{\partial^3 T}{\partial x^3}|_{x_k} + \mathcal{O}(\Delta x^4), \tag{15.15}$$

$$T_{k-1}^n = T_k^n - \frac{\partial T}{\partial x}|_{x_k}\Delta x + \frac{\Delta x^2}{2}\frac{\partial^2 T}{\partial x^2}|_{x_k} - \frac{\Delta x^3}{6}\frac{\partial^3 T}{\partial x^3}|_{x_k} + \mathcal{O}(\Delta x^4), \tag{15.16}$$

which summed together give

$$\frac{T_{k+1}^n - 2T_k^n + T_{k-1}^n}{\Delta x^2} = \frac{\partial^2 T}{\partial x^2} + \underbrace{\frac{\partial^4 T}{\partial x^4}\frac{\Delta x^2}{12} + \mathcal{O}(\Delta x^4)}_{\textit{truncation in space } \mathcal{T}_x}. \tag{15.17}$$

The FDE can then be related to the PDE

$$\frac{\partial T}{\partial t} + \underbrace{\frac{\Delta t}{2}\frac{\partial^2 T}{\partial t^2} + \mathcal{O}(\Delta t^2)}_{\mathcal{T}_t} = D\frac{\partial^2 T}{\partial x^2} + \underbrace{D\left(\frac{\partial^4 T}{\partial x^4}\frac{\Delta x^2}{12} + \mathcal{O}(\Delta x^4)\right)}_{\mathcal{T}_x}. \tag{15.18}$$

The explicit scheme is consistent as both $\mathcal{T}_t$ and $\mathcal{T}_x \rightarrow 0$ when $\Delta t, \Delta x \rightarrow 0$. Moreover, the explicit scheme is first order in time and second order in space.

15.2.4 **Stability**

This is the first application of the von Neumann stability method discussed in Chapter 13. The main strategy here is to decompose the solution to the FDE in terms of an infinite series

$$T_k^n = \sum_j a_j^n g_j(x_k), \tag{15.19}$$

where the time dependence is described by the amplitude coefficients $a_j(t)$ and the spatial dependence by the functions $g_j(x)$. Introducing this form into the FDE allows us to compare the amplitude or weight of each term in the series (the magnitude of the a_js) at successive time steps to establish if they grow, remain stable, or shrink over time. If we can establish that for all possible choices of j and for any successive time steps n and $n+1$, the magnitude of the a_js does not increase for a choice of Δt and Δx, then the scheme is stable under these conditions.

We show here how this method can be applied to test the stability of the explicit diffusion scheme. It is convenient to use sine test functions for g_j so that

$$g_j(x_k) = \sin(j\pi k\Delta x). \tag{15.20}$$

The numerical solution for the temperature at position $x_k = k\Delta x$ and time $t_n = n\Delta t$ is therefore

$$T_k^n = \sum_j a_j^n \sin(j\pi k\Delta x). \tag{15.21}$$

This expression can be introduced into the FDE to get

$$\sum_j \left(a_j^{n+1} - a_j^j\right) \sin(j\pi k\Delta x) = \underbrace{\frac{D\Delta t}{\Delta x^2}}_{\mathcal{D}} \sum_j a_j^n \left[\sin(j\pi(k+1)\Delta x) - 2\sin(j\pi k\Delta x) + \sin(j\pi(k-1)\Delta x)\right]. \tag{15.22}$$

We can now use the trigonometric property

$$\sin(\alpha)\cos(\beta) = \frac{1}{2}\left[\sin(\alpha-\beta) + \sin(\alpha+\beta)\right], \tag{15.23}$$

with $\alpha = j\pi k\Delta x$ and $\beta = j\pi\Delta x$, which allows us to recast some of the terms in the right-hand side of equation 15.22 as

$$\sin(j\pi(k+1)\Delta x) + \sin(j\pi(k-1)\Delta x) = 2\sin(j\pi k\Delta x)\cos(j\pi\Delta x). \tag{15.24}$$

It is now possible to obtain an expression for the FDE in terms of factors of $\sin(j\pi k\Delta x)$

$$\sum_j \left(a_j^{n+1} - a_j^j\right) \sin(j\pi k\Delta x) = \mathcal{D}\sum_j a_j^n \sin(j\pi k\Delta x)\left[2\cos(j\pi\Delta x) - 2\right]. \tag{15.25}$$

We introduce the parameter $\sigma_j \equiv 2(1-\cos(j\pi\Delta x))$, which depends on k but is bound ($0 \le \sigma_j \le 4$), which will become important later. With this definition our FDE becomes

$$\sum_j \left(a_j^{n+1} - a_j^j\right) \sin(j\pi k\Delta x) = -\mathcal{D}\sum_j a_j^n \sin(j\pi k\Delta x)\sigma_j. \tag{15.26}$$

It is time now to use the orthogonality of the family of functions $\sin(j\pi x)$ (as seen in Part I, section 4)

$$\int_0^1 \sin(j\pi x)\sin(m\pi x)dx = \begin{cases} \frac{1}{2} & \text{if } j = m, \\ 0 & \text{otherwise.} \end{cases} \quad (15.27)$$

Remembering that $k\Delta x$ is the position x (at point x_k), we can multiply each side of the FDE by $\sin(m\pi x)$ and integrate over the interval $0 \leq x \leq 1$

$$\sum_j \left(a_j^{n+1} - a_j^j\right) \int_0^1 \sin(j\pi x)\sin(m\pi x)dx = -\mathcal{D}\sum_j a_j^n \sigma_j \int_0^1 \sin(j\pi x)\sin(m\pi x)dx. \quad (15.28)$$

Using the orthogonality condition above allows us to reduce the sums to a single term with coefficients a_m^n and a_m^{n+1}, which enables us to find conditions over which the amplitude associated with the element of solution for T with the arbitrary wavenumber m (i.e., the condition would be true for any value of m)

$$a_m^{n+1} - a_m^n = -\mathcal{D}\sigma_m a_m^n, \quad (15.29)$$

which we can reorganize to get

$$\left|\frac{a_m^{n+1}}{a_m^n}\right| = |1 - \mathcal{D}\sigma_m|. \quad (15.30)$$

For the amplitude of the mode m of the solution to remain bounded for any number of time steps, this expression should be such that

$$|1 - \mathcal{D}\sigma_m| \leq 1, \quad (15.31)$$

which implies two conditions First,

$$\mathcal{D}\sigma_m \geq 0, \quad (15.32)$$

which given the positivity of $\mathcal{D}$ and σ_m is a trivial condition that is always satisfied. The second condition is more interesting and brings

$$-1 \leq 1 - \mathcal{D}\sigma_m, \quad (15.33)$$

which using a conservative approximation for $\sigma_m = 4$ (maximum possible value) yields a condition on the diffusion number for the explicit scheme that is independent of the mode considered (value of m)

$$\mathcal{D} = \frac{D\Delta t}{\Delta x^2} \leq \frac{1}{2}. \quad (15.34)$$

The scheme is therefore conditionally stable and the choice of time step to satisfy absolute stability is

$$\Delta t \leq \frac{\Delta x^2}{2D}. \quad (15.35)$$

15.2.5 **Convergence**

The Lax–Richtmyer equivalence theorem states that consistency and stability are necessary and sufficient to establish convergence. The explicit scheme for the diffusion equation is therefore convergent for $\mathcal{D} \leq 1/2$. This explains why the results of our calculations in the right side of Figure 15.2 were unstable ($\mathcal{D} = 0.6$ in that example).

The conditional stability of the explicit scheme can be limiting, as it reduces the maximum time step that the scheme is allowed to use to integrate over time. It is therefore important to introduce two additional models: the implicit and the Crank–Nicolson schemes.

15.3 **Scheme 2: Implicit (BE/time) approach**

The idea of the implicit scheme is to change the time at which the function T is evaluated in the spatial derivative term. The BE approach in time is used to retrieve a (set of) FDE(s)

$$\frac{T_k^{n+1} - T_k^n}{\Delta t} = D\frac{T_{k+1}^{n+1} - 2T_k^{n+1} + T_{k-1}^{n+1}}{\Delta x^2}. \tag{15.36}$$

15.3.1 **Implementation**

Although the difference between the explicit and implicit schemes seems subtle, the impact of that change is quite profound in two major ways. First, the implementation is more complex and requires more effort. Second, the improvement in terms of stability make it worth the effort. The stencil of the implicit method is shown in Figure 15.3 and we can already see why implementation is becoming more challenging, as each algebraic equation now contains three unknowns T_k^{n+1} as does the explicit scheme, as well as T_{k-1}^{n+1} and T_{k+1}^{n+1}. Retrieving a marching equation therefore requires more care. The challenge presented to us is however not new; we dealt with a similar problem when we studied linear box models and more specifically how to implement a BE scheme to solve them (see Part 9).

First, we recast the FDE to isolate the three unknowns on the left-hand side

$$-\mathcal{D}T_{k+1}^{n+1} + (1 + 2\mathcal{D})\,T_k^{n+1} - \mathcal{D}T_{k-1}^{n+1} = T_k^n. \tag{15.37}$$

This equation is true for all values of k within our domain (with some subtle modifications to account for BCs). Ordering the Ts in terms of a vector spanning all the the positions within the domain ($k = 1, 2, \ldots Nx$)

$$\mathbf{T}^n = \begin{pmatrix} T_1^n \\ T_2^n \\ \vdots \\ T_{Nx}^n \end{pmatrix}, \tag{15.38}$$

we can build a marching equation of the form

$$\mathbf{M}\mathbf{T}^{n+1} = \mathbf{T}^n, \tag{15.39}$$

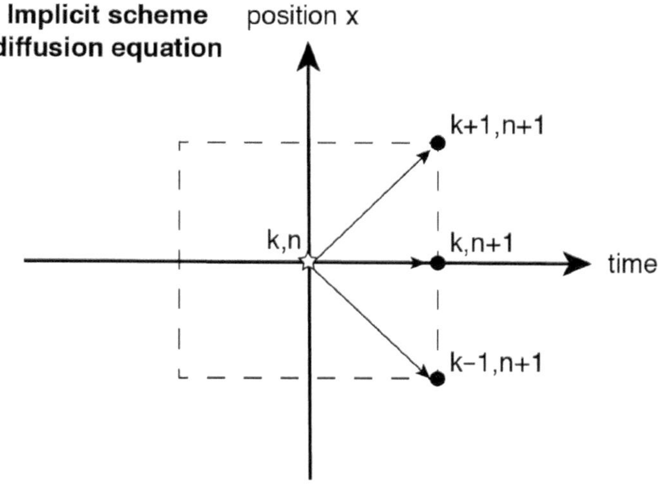

Figure 15.3 Stencil for the implicit (backward in time) scheme for the diffusion equation.

with a matrix **M** with dimensions $Nx \times Nx$

$$\mathbf{M} = \begin{pmatrix} 1+2D & -D & 0 & \dots & & & \\ -D & 1+2D & -D & 0 & \dots & & \\ 0 & -D & 1+2D & -D & 0 & \dots & \\ \vdots & & & & & & \\ & & \dots & 0 & -D & 1+2D & \\ & & & & & & \end{pmatrix}. \tag{15.40}$$

M is a tridiagonal matrix and, for large domains (large values of Nx), it is sparse and can be inverted efficiently with the Thomas algorithm (Section 15.3.2). The marching equation at each time step for the implicit scheme is therefore

$$\mathbf{T}^{n+1} = \underbrace{\mathbf{M}^{-1}}_{\equiv \mathbf{M}_{BE}} \mathbf{T}^{n}. \tag{15.41}$$

The set of marching equations in (Eq. 15.39) is not fully consistent with the FDE in (Eq. 15.37). This is because of the first and last row of the vectorial system of equations. The FDE for $k = 1$ and $k = Nx$ reads

$$-DT_2^{n+1} + (1+2D)\,T_2^{n+1} - DT_0^{n+1} = T_1^n, \tag{15.42}$$

$$-DT_{Nx+1}^{n+1} + (1+2D)\,T_{Nx}^{n+1} - DT_{Nx-1}^{n+1} = T_{Nx}^n, \tag{15.43}$$

while the vectorial marching equation implies

$$-\mathcal{D}T_2^{n+1} + (1 + 2\mathcal{D})\, T_2^{n+1} = T_1^n, \tag{15.44}$$

$$(1 + 2\mathcal{D})\, T_{Nx}^{n+1} - \mathcal{D}T_{Nx-1}^{n+1} = T_{Nx}^n. \tag{15.45}$$

The latter expressions omit the terms depending on T_0^{n+1} and T_{Nx+1}^{n+1}. It is worth thinking about what these terms represent before moving further. The diffusion equation admits two BCs (because of its second-order spatial derivative), T_0 and T_{Nx+1} are outside the domain of calculations and actually relate to these two BCs. Let us consider two generic cases, the first with Dirichlet BCs (temperature T imposed at the boundary in our example) and the second with Neuman BCs (imposed heat flux or $\frac{\partial T}{\partial x}$ at the two boundaries). In the first case (imposed values for $T_0 = T_a$ and $T_{Nx+1} = T_b$, as the BCs are known and imposed, we can just reformulate the two equations above as

$$-\mathcal{D}T_2^{n+1} + (1 + 2\mathcal{D})\, T_2^{n+1} = T_1^n + \mathcal{D}T_a. \tag{15.46}$$

$$(1 + 2\mathcal{D})\, T_{Nx}^{n+1} - \mathcal{D}T_{Nx-1}^{n+1} = T_{Nx}^n + \mathcal{D}T_b. \tag{15.47}$$

In essence, we can use the vectorial marching equation provided the right-hand side vector is amended

$$\mathbf{T}^{n+1} = \underbrace{\mathbf{M}^{-1}}_{\equiv \mathbf{M}_{BE}} \hat{\mathbf{T}}^n, \tag{15.48}$$

where

$$\hat{\mathbf{T}}^n = \mathbf{T}^n + \mathbf{V}_{BC} = \begin{pmatrix} T_1^n \\ T_2^n \\ \vdots \\ T_{Nx}^n \end{pmatrix} + \begin{pmatrix} \mathcal{D}T_a \\ 0 \\ \vdots \\ \mathcal{D}T_b. \end{pmatrix} = \begin{pmatrix} T_1^n + \mathcal{D}T_a \\ T_2^n \\ \vdots \\ T_{Nx}^n + \mathcal{D}T_b. \end{pmatrix} \tag{15.49}$$

The exercise is somewhat more difficult for flux BCs. We first need to manipulate the FDEs for $k = 1$ and $k = Nx$ to highlight how fluxes are implicitly hidden in the FDEs. The generic FDE for the implicit scheme can be written in the form

$$\frac{T_k^{n+1} - T_k^n}{\Delta t} = \frac{D}{\Delta x}\left[\left(\frac{T_{k+1}^{n+1} - T_k^{n+1}}{\Delta x}\right) - \left(\frac{T_k^{n+1} - T_{k-1}^{n+1}}{\Delta x}\right)\right], \tag{15.50}$$

where the terms

$$\frac{T_{k+1}^{n+1} - T_k^{n+1}}{\Delta x} \quad \text{and} \quad \frac{T_k^{n+1} - T_{k-1}^{n+1}}{\Delta x} \tag{15.51}$$

relate to $\partial T/\partial x$ going out of and coming into the node k respectively. In the case of a flux or Neuman BC, $f_{left} = \partial T/\partial x|_{left}$ and $f_{right} = \partial T/\partial x|_{right}$ are specified at the boundary. The FDEs for $k = 1$ and $k = Nx$ are then modified to introduce these BCs accordingly

$$\frac{T_1^{n+1} - T_1^n}{\Delta t} = \frac{D}{\Delta x}\left[\left(\frac{T_2^{n+1} - T_1^{n+1}}{\Delta x}\right) - f_{left}\right], \tag{15.52}$$

$$\frac{T_{Nx}^{n+1} - T_{Nx}^n}{\Delta t} = \frac{D}{\Delta x}\left[f_{right} - \left(\frac{T_{Nx}^{n+1} - T_{Nx-1}^{n+1}}{\Delta x}\right)\right]. \tag{15.53}$$

The choice of BC therefore affects the definition of the matrix-vector marching equation (Eq. 15.48)

$$\mathbf{T}^{n+1} = \underbrace{\mathbf{M}'^{-1}}_{\equiv \mathbf{M}_{BE'}} \tilde{\mathbf{T}}^{n}, \tag{15.54}$$

where

$$\mathbf{M}' = \begin{pmatrix} 1+\mathcal{D} & -\mathcal{D} & 0 & \dots & & & \\ -\mathcal{D} & 1+2\mathcal{D} & -\mathcal{D} & 0 & \dots & & \\ 0 & -\mathcal{D} & 1+2\mathcal{D} & -\mathcal{D} & 0 & \dots & \\ \vdots & & & & & & \\ & & \dots & 0 & -\mathcal{D} & 1+\mathcal{D} & \end{pmatrix} \tag{15.55}$$

and

$$\tilde{\mathbf{T}}^{n} = \begin{pmatrix} T_1^n - \mathcal{D}\Delta x f_{left} \\ T_2^n \\ \vdots \\ T_{Nx}^n + \mathcal{D}\Delta x f_{right}. \end{pmatrix} \tag{15.56}$$

A pseudo-code for the implementation of the implicit scheme is provided below. There are several ways we can proceed to invert the matrix $\mathbf{M}$ or $\mathbf{M}'$. It is important though to realize that the $Nx \times Nx$ matrix is sparse (mostly 0s with only nonzero terms in the diagonal, lower diagonal, and upper diagonal). There is a very efficient way to proceed when inverting these tridiagonal matrices which is worth the effort; it is sometimes called the Thomas algorithm or tridiagonal matrix inversion (TDMA). Note that using the TDMA requires small changes in the pseudo-code as the inversion process is now moved into the time integration loop, although it remains more efficient than the pseudo-code described here.

```
dt=...;
dx=...;
D=...;
Diff_n=D*dt/dx^2;
T_old=...; % Initial condition
M=... % build matrix M or M'
M_BE = ... % inversion of M or M'
for time=1:maxtime-1 % Marching equation
        RHS=T_old+V_BC; % modify RHS with BC
        T=M_BE *RHS;
        T_old=T;
end
```

15.3.2 **Tri-diagonal matrix solver, or Thomas algorithm (TDMA)**

The TDMA is designed to find the solution to

$$\mathbf{M}\mathbf{x} = \mathbf{y}, \tag{15.57}$$

where $\mathbf{M}$ is a tridiagonal matrix and $\mathbf{x}$ and $\mathbf{y}$ are two vectors of same dimension (i.e., $\mathbf{M}$ is a square matrix). For simplicity, we define

$$\mathbf{M} = \begin{pmatrix} d_1 & u_1 & 0 & \dots & & & \\ l_2 & d_2 & u_2 & 0 & \dots & & \\ 0 & l_3 & d_3 & u_3 & 0 & \dots & \\ \vdots & & & & & & \\ & & \dots & 0 & l_N & d_N \end{pmatrix} \tag{15.57}$$

and by definition the elements $l_1 = u_N = 0$. The idea is to solve for $\mathbf{x}$ with a forward and backward substitution.

The first step is to build the forward substitution. To that end, we write the first algebraic equation

$$d_1 x_1 + u_1 x_2 = y_1. \tag{15.59}$$

We can isolate the first element of the vector $\mathbf{x}$

$$x_1 = \frac{y_1 - u_1 x_2}{d_1}. \tag{15.60}$$

The second equation reads

$$l_2 x_1 + d_2 x_2 + u_2 x_3 = y_2, \tag{15.61}$$

and using (Eq. 15.60) to eliminate x_1 in this equation yields

$$\underbrace{\left(d_2 - \frac{l_2 u_1}{d_1}\right)}_{P_2} x_2 + u_2 x_3 = \underbrace{y_2 - \frac{l_2 y_1}{d_1}}_{Q_2}, \tag{15.62}$$

where we define $m_j = l_j / d_{j-1}$ in

$$Q_2 = y_2 - m_2 y_1, \tag{15.63}$$

and

$$P_2 = d_2 - m_2 u_1. \tag{15.64}$$

Now, addressing the third equation and eliminating x_2 we get

$$\left(d_3 - \frac{u_2 l_3}{P_2}\right) x_3 + u_3 x_4 = \underbrace{y_3 - \frac{l_3 Q_2}{P_2}}_{Q_3}. \tag{15.65}$$

The same process of forward substitution repeats and we can easily show that we obtain recursive expressions

$$m_k = \frac{l_k}{P_{k-1}} \tag{15.66}$$

$$P_k = d_k - m_k u_{k-1} \tag{15.67}$$

$$Q_k = y_k - m_k Q_{k-1}. \tag{15.68}$$

Finally, for the last equation (row N)

$$P_N x_N = Q_N, \tag{15.69}$$

which gives

$$x_N = \frac{Q_N}{P_N}. \tag{15.70}$$

In essence, the forward substitution serves to build the P_j and Q_j that we use in the backward substitution to get the element of the vector **x**. Now we can start the backward substitution as we already know x_N and can retrieve x_{N-1} from

$$P_{N-1}x_{N-1} + u_{N-1}x_N = Q_{N-1}, \tag{15.71}$$

which gives

$$x_{N-1} = \frac{Q_{N-1} - u_{N-1}x_N}{P_{N-1}}, \tag{15.72}$$

which allows us to compute x_{N-2}

$$x_{N-2} = \frac{Q_{N-2} - u_{N-2}x_{N-1}}{P_{N-2}}, \tag{15.73}$$

and so forth until

$$x_1 = \frac{Q_1 - u_1 X_2}{P_1}. \tag{15.74}$$

The number of operations required to compute the inverse and solve the linear system is order of N here instead of N^3 with a brute force inversion method (like Gaussian elimination). It therefore becomes quickly advantageous, especially when considering large domains or high spatial resolution (high N).

15.3.3 Performance

We repeat the same numerical experiment that we did for the explicit scheme and find the results shown in Figure 15.4. There is again a good match with the analytical solution for the run conducted at $D = 0.2$. The main difference here is that the implicit scheme seems to be stable at $D = 0.6$, which was not true for the explicit scheme. The cost of the tridiagonal matrix inversion seems to be offset by the fact that the implicit model is more stable. The question is how stable? Before answering this question, we discuss the consistency of the implicit scheme.

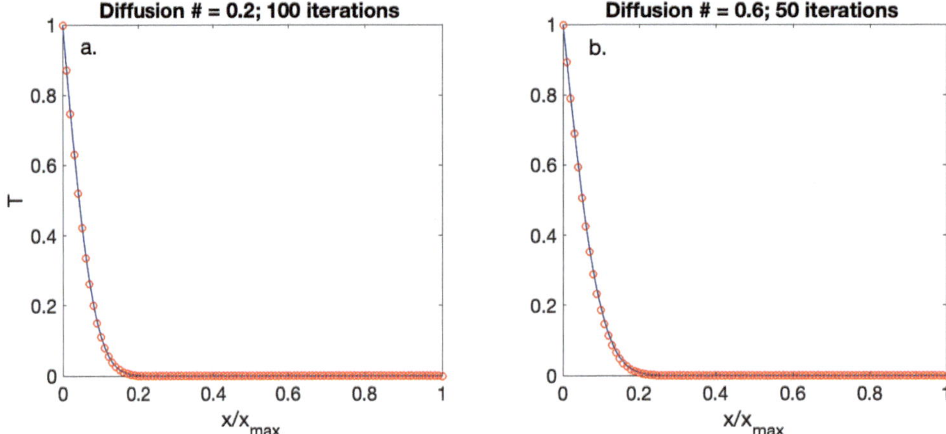

Figure 15.4 Two realizations of the implicit scheme for diffusion. a): The diffusion number $\mathcal{D} = 0.2$ and the simulation is run for 100 iterations. b): The diffusion number is increased to 0.6 and the simulation covers only 50 iterations. In contrast with the explicit scheme, the implicit scheme is stable.

15.3.4 Consistency

The derivation of the consistency of the explicit scheme can be repeated here. The only difference is that the spatial derivatives of the function T are taken at the time $t + \Delta t$ instead of t, and the rest remains identical. The implicit scheme for the diffusion equation is therefore consistent: it is first order in time and second order in space similarly to the explicit scheme.

15.3.5 Stability

Establishing the stability condition for the implicit scheme follows closely from the work we have done with the explicit model. We assume that the numerical solution at position x_k and time $n\Delta t$ can be expressed as

$$T_k^n = \sum_j a_j^n \sin(j\pi k\Delta x). \tag{15.75}$$

Plugging this expression back into the FDE brings

$$\sum_j \left(a_j^{n+1} - a_j^j\right) \sin(j\pi k\Delta x) = \underbrace{\frac{D\Delta t}{\Delta x^2}}_{\mathcal{D}} \sum_j a_j^{n+1} \left[\sin((k+1)\pi j\Delta x) - 2\sin(j\pi k\Delta x) + \sin((k-1)\pi j\Delta x)\right]. \tag{15.76}$$

Using the same trigonometric property and the orthogonality of the sin function we retrieve a different equation for the amplitudes a_m

$$a_m^{n+1} - a_m^n = -\mathcal{D}\sigma_m a_m^{n+1}, \tag{15.77}$$

where $\sigma_m \equiv 2(1 - 2\cos(m\pi\Delta x))$, which we can reorganize to get

$$\left|\frac{a_m^{n+1}}{a_m^n}\right| = \left|\frac{1}{1 + \mathcal{D}\sigma_m}\right| < 1, \tag{15.78}$$

which, given the positivity of D and σ_m, is a trivial condition that is always satisfied. The implicit scheme is therefore unconditionally stable. The subtle change between explicit and implicit schemes has therefore a profound effect on the performance of the two models for the diffusion equation.

15.3.6 **Convergence**

The implicit scheme is unconditionally convergent (consistent and unconditionally stable), which makes it the standard numerical method to solve diffusion problems numerically. The computational cost (and effort) associated with the build-up and inversion of the tridiagonal matrix is largely offset by the possibility to increase the time step without fearing for numerical instabilities. Perhaps one limitation of the implicit scheme is its accuracy in time, as the scheme is first order. Section 15.4 introduces the Crank–Nicolson scheme for the diffusion equation. In a crude way, it can be seen as a combination of explicit and implicit schemes, but it preserves the unconditional stability of the implicit scheme with now a second-order truncation error in time. It is a widely used scheme as well because of its performance.

15.4 **Scheme 3: Crank–Nicolson approach**

The main goal of scheme 3 is to preserve the unconditional stability while improving the temporal accuracy of the implicit model. The idea is simple: the finite difference is treated as a combination of the explicit and implicit models with equal weights (1/2)

$$\frac{T_k^{n+1} - T_k^n}{\Delta t} = \frac{D}{2\Delta x^2}\left(T_{k+1}^n - 2T_{k-1}^n + T_{k-1}^n\right) + \frac{D}{2\Delta x^2}\left(T_{k+1}^{n+1} - 2T_k^{n+1} + T_{k-1}^{n+1}\right). \tag{15.79}$$

15.4.1 **Implementation**

Similarly to the implicit scheme, each algebraic FDE contains three unknowns. The best way to codify these equations is therefore a matrix-vector system of FDEs. For conciseness, we use a case with Dirichlet BCs T_a and T_b.

$$\mathbf{T}^{n+1} = \underbrace{\mathbf{M}^{-1}}_{\equiv \mathbf{M}_{CN}} \mathbf{R}, \tag{15.80}$$

where

$$\mathbf{M} = \begin{pmatrix} 1+\mathcal{D} & -\frac{\mathcal{D}}{2} & 0 & \dots & & & \\ -\frac{\mathcal{D}}{2} & 1+\mathcal{D} & -\frac{\mathcal{D}}{2} & 0 & \dots & & \\ 0 & -\frac{\mathcal{D}}{2} & 1+\mathcal{D} & -\frac{\mathcal{D}}{2} & 0 & \dots & \\ \vdots & & & & & & \\ & & \dots & 0 & -\frac{\mathcal{D}}{2} & 1+\mathcal{D} & \end{pmatrix} \tag{15.81}$$

and

$$\mathbf{R} = \begin{pmatrix} T_1^n(1-\mathcal{D}) + \mathcal{D}T_a + \frac{\mathcal{D}}{2}T_2^n \\ T_2^n(1-\mathcal{D}) + \frac{\mathcal{D}}{2}T_1^n + \frac{\mathcal{D}}{2}T_3^n \\ \vdots \\ T_{Nx}^n\left(1-\frac{\mathcal{D}}{2}\right) + \frac{\mathcal{D}}{2}T_{Nx-1}^n + \mathcal{D}T_b \end{pmatrix}. \tag{15.82}$$

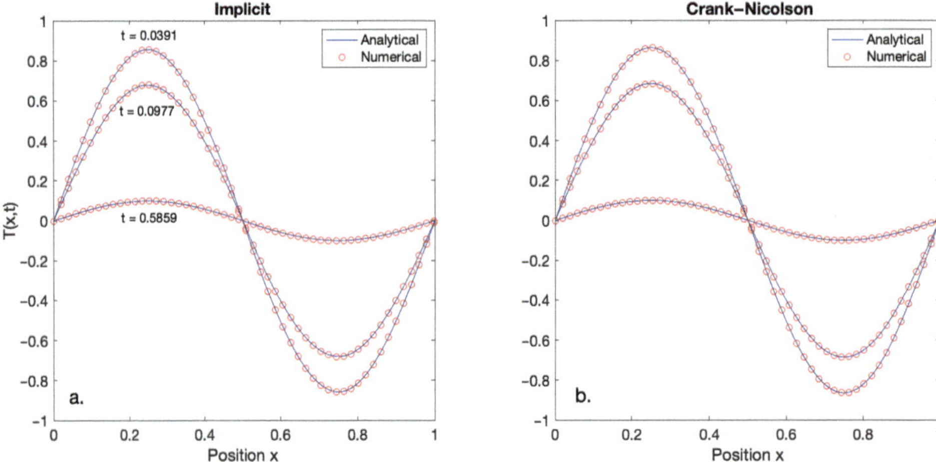

Figure 15.5 Comparison between the implicit scheme and the analytical solution at different times (left panel). Same simulations with the Crank–Nicolson scheme (right panel).

The matrix **M** is again tridiagonal and we can apply the TDMA to solve for the function T at each time step.

15.4.2 Performance

We test this algorithm with a new test problem that has a well-known analytical solution. We assume an initial condition $T(x, 0) = \sin(\pi kx)$ with boundaries at $x = 0, 1$ where $T = 0$. The analytical solution is

$$T(x, t) = sin(\pi kx) \exp(-D(k\pi)^2 t), \tag{15.83}$$

where D is the diffusion coefficient. Figure 15.5 shows a few snapshots of the numerical solutions with the implicit and Crank–Nicolson schemes compared with the analytical solution. In these two simulations, 512 points were used $dx = 0.002$ and $dt = 0.0078$ and $D = 0.1$.

Another analysis is warranted to observe the benefit of using the Crank–Nicolson scheme. We fix the grid spacing to $dx = 0.002$, conduct a series of diffusion simulations covering each a total time of $t = 2$ but consider a range of number of time steps from 16 ($dt = 0.125$) up to 1024 ($dt = 0.002$). We compute for each run the error of the scheme compared to the analytical solution at the end time

$$Error = \frac{\sum_{k=1}^{512} \left(T_k^{t=\text{end}} - T(x_k, t = \text{end})\right)^2}{\sqrt{512}}. \tag{15.84}$$

The objective of this test is to determine the accuracy in time (associated with the truncation error in time) of each scheme.

15.4.3 Consistency

The finite difference scheme involves several terms that approximate $T(x, t)$ at positions and times that differ from x_k and t_n respectively. Using Taylor expansions to various orders we can develop

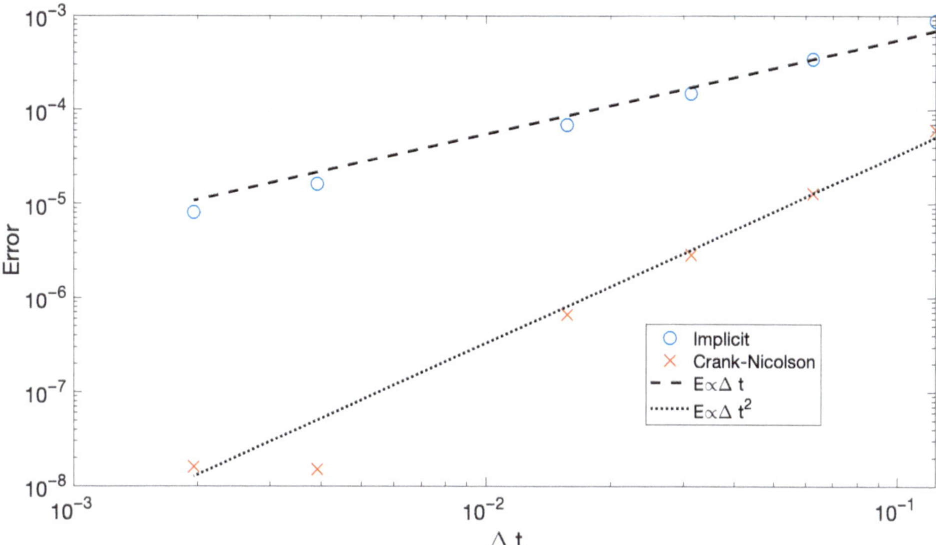

Figure 15.6 Error in time for the implicit and the Crank–Nicolson schemes. Note that the convergence rate is first order of implicit and second order for the Crank–Nicolson as predicted by the truncation error (see Section 15.4.3).

each component of the equation in terms of T_k^n

$$T_k^{n+1} = T_k^n + \frac{\partial T}{\partial t}\Delta t + \frac{\Delta t^2}{2}\frac{\partial^2 T}{\partial t^2} + \mathcal{O}(\Delta t^3) \tag{15.85}$$

$$T_{k+1}^{n} = T_k^n + \frac{\partial T}{\partial x}\Delta x + \frac{\Delta x^2}{2}\frac{\partial^2 T}{\partial x^2} + \mathcal{O}(\Delta x^3) \tag{15.86}$$

$$T_{k-1}^{n} = T_k^n - \frac{\partial T}{\partial x}\Delta x + \frac{\Delta x^2}{2}\frac{\partial^2 T}{\partial x^2} + \mathcal{O}(\Delta x^3) \tag{15.87}$$

$$T_{k+1}^{n+1} = T_k^n + \frac{\partial T}{\partial t}\Delta t + \frac{\partial T}{\partial x}\Delta x + \frac{\Delta t^2}{2}\frac{\partial^2 T}{\partial t^2} + \frac{\Delta x^2}{2}\frac{\partial^2 T}{\partial x^2} + \frac{\Delta t\Delta x}{2}\frac{\partial^2 T}{\partial t\partial x} + \frac{\Delta t\Delta x^2}{2}\frac{\partial^3 T}{\partial t\partial x^2} + \mathcal{O}(\Delta x^3, \Delta t^3) \tag{15.88}$$

$$T_{k-1}^{n+1} = T_k^n + \frac{\partial T}{\partial t}\Delta t - \frac{\partial T}{\partial x}\Delta x + \frac{\Delta t^2}{2}\frac{\partial^2 T}{\partial t^2} + \frac{\Delta x^2}{2}\frac{\partial^2 T}{\partial x^2} - \frac{\Delta t\Delta x}{2}\frac{\partial^2 T}{\partial t\partial x} + \frac{\Delta t\Delta x^2}{2}\frac{\partial^3 T}{\partial t\partial x^2} + \mathcal{O}(\Delta x^3, \Delta t^3). \tag{15.89}$$

Substituting each term in the finite difference and grouping them in terms of powers of Δt yields

$$\underbrace{\left(\frac{\partial T}{\partial t} - D\frac{\partial^2 T}{\partial x^2}\right)}_{=0 \quad PDE} + \underbrace{\frac{\Delta t}{2}\left(\frac{\partial^2 T}{\partial t^2} - D\frac{\partial^3 T}{\partial t\partial x^2}\right) + \mathcal{O}(\Delta x^2, \Delta t^2)}_{Truncation} = 0. \tag{15.90}$$

We can reorganize the truncation error term and get

$$\frac{\Delta t}{2}\left(\frac{\partial^2 T}{\partial t^2} - D\frac{\partial^3 T}{\partial t \partial x^2}\right) = \frac{\Delta t}{2}\frac{\partial}{\partial t}\underbrace{\left(\frac{\partial T}{\partial t} - D\frac{\partial^2 T}{\partial x^2}\right)}_{=0 \quad PDE}, \tag{15.91}$$

which shows that the Crank–Nicolson scheme for the diffusion equation is second order in space **and** time.

15.4.4 **Stability**

The same approach applies to the Crank–Nicolson method when testing its stability. We assume the numerical solution can be cast in terms of a sin series

$$T_k^n = \sum_j a_j^n \sin(j\pi k\Delta x). \tag{15.92}$$

Plugging this expression back into the FDE brings

$$\sum_j \left(a_j^{n+1} - a_j^j\right) \sin\left(j\pi k\Delta x\right) = \underbrace{\frac{D\Delta t}{\Delta x^2}}_{\mathcal{D}} \sum_j \frac{a_j^{n+1} - a_j^n}{2}$$

$$\left[\sin((k+1)\pi j\Delta x) - 2\sin(j\pi k\Delta x) + \sin((k-1)\pi j\Delta x)\right]. \tag{15.93}$$

Using the same trigonometric property and the orthogonality of the sin function we retrieve now a different equation for the amplitudes a_m

$$a_m^{n+1} - a_m^n = -\frac{\mathcal{D}}{2}\sigma_m\left(a_m^{n+1} + a_m^n\right), \tag{15.94}$$

where $\sigma_m \equiv 2(1 - 2\cos(m\pi\Delta x))$. We can reorganize this expression to get

$$\left|\frac{a_m^{n+1}}{a_m^n}\right| = \left|\frac{1 - \frac{\mathcal{D}}{2}\sigma_m}{1 + \frac{\mathcal{D}}{2}\sigma_m}\right|. \tag{15.95}$$

The two conditions for stability are then

$$-\left(1 + \frac{\mathcal{D}\sigma_m}{2}\right) \leq 1 - \frac{\mathcal{D}\sigma_m}{2} \leq 1 + \frac{\mathcal{D}\sigma_m}{2}. \tag{15.96}$$

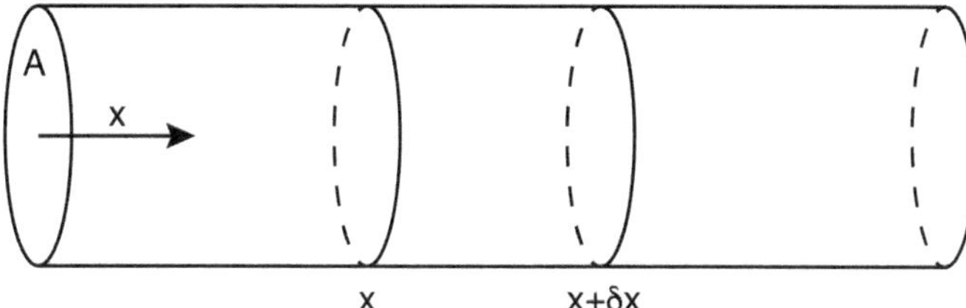

Figure 15.7 One-dimensional diffusion setup for the derivation of our energy conservation equation.

Using that $0 \le \sigma_m \le 4$, the condition on the right is trivial and always satisfied if $D > 0$, and the condition on the left is also always satisfied (leads to $-2 \le 0$). The Crank–Nicolson scheme is therefore unconditionally stable.

15.4.5 **Convergence**

The scheme has preserved the stability (unconditional) of the implicit method, but the truncation error in time is now improved (second order). The implementation is more demanding than even the implicit scheme, although both rely on the inversion of a tridiagonal matrix and would be roughly similar in terms of computational cost. Importantly though, the accuracy is improved, which combined to the unconditional stability offers the opportunity to run simulations with relatively large time steps without incurring the heavy cost of a first-order scheme in time.

15.5 **Melting/solidification: The Stefan problem**

Heat transfer with phase change is common to many applications in science and engineering. One example for us would be a simple model to compute the melting rate of an ice sheet caused by volcanic activity, the freezing time scale of sea ice, or the solidification rate inside dikes to cite a few. In each of these problems, the latent heat of fusion becomes an important part of the energy balance. Nevertheless, the simple energy balance model we discussed when introducing the concept of numerical diffusion can be a decent starting point.

Assuming a one-dimensional domain (depicted in Figure 15.7), we consider now the conservation of the enthalpy per unit volume $h(x, t)$. For simplicity we assume that the density and other material properties of the substance considered are constant. Furthermore, we assume that the material is a pure substance (liquidus temperature = solidus temperature), such as water and ice. The melt fraction $0 \le f(x, t) \le 1$ describes the volume fraction of melt present locally. In the section of the material considered between positions x and $x + \delta x$, we denote the average enthalpy at time t within that section to be

$$h\left(x+\frac{\delta x}{2},t\right)A\delta x = \rho cT\left(x+\frac{\delta x}{2},t\right) + \rho L_f f\left(x+\frac{\delta x}{2},t\right)A\delta x, \tag{15.97}$$

where we considered the half-point between x and $x + \delta x$ to define the average temperature and melt fraction arbitrarily. Now, considering two instances of time separated by an interval δt, the change in enthalpy in the section over that interval is

$$\delta h = \rho c\left(T\left(x + \frac{\delta x}{2}, t + \delta t\right) - T\left(x + \frac{\delta x}{2}, t\right)\right) + \rho L_f\left(f\left(x + \frac{\delta x}{2}, t + \delta t\right) - f\left(x + \frac{\delta x}{2}, t\right)\right) A\delta x. \quad (15.98)$$

This change of enthalpy is balanced by the energy coming into and out of the section over the same time interval

$$\delta h = -\left(q\left(x + \delta x, \frac{\delta t}{2}\right) - q\left(x, \frac{\delta t}{2}\right)\right) A\delta t. \quad (15.99)$$

Dividing both sides by $A\delta t\delta x$ yields

$$\frac{\rho c\left(T\left(x + \frac{\delta x}{2}, t + \delta t\right) - T\left(x + \frac{\delta x}{2}, t\right)\right)}{\delta t} = -\frac{\left(q\left(x + \delta x, \frac{\delta t}{2}\right) - q\left(x, \frac{\delta t}{2}\right)\right)}{\delta x} - \rho L_f \frac{\left(f\left(x + \frac{\delta x}{2}, t + \delta t\right) - f\left(x + \frac{\delta x}{2}, t\right)\right)}{\delta t}. \quad (15.100)$$

Taking the limit of $\delta t, \delta x \to 0$ and assuming a homogeneous thermal conductivity k, we can use the definition of $q = -k\partial T/\partial x$ and get the PDE for the thermal budget associated with melting of a pure substance in one dimension

$$\frac{\partial T}{\partial t} = \underbrace{\frac{k}{\rho c}}_{\kappa} \frac{\partial^2 T}{\partial x^2} - \frac{L}{c}\frac{\partial f}{\partial t}, \quad (15.101)$$

where κ is the thermal diffusivity of the substance. This equation is conceptually problematic: we have two scalar fields to solve for $T(x, t)$ and $f(x, t)$, but only one equation. For closure, we need an equation that relates the melt fraction f to the temperature field T, which for a pure substance is again problematic, given that f can be anything between 0 and 1 at the melting temperature T_m. That issue introduces some circularity in our solution procedure. Given the relationship between T and f is not continuous (not bijective would be a better word), enthalpy is a better suited quantity to relate to f. Defining the maximum enthalpy for a solid to be $h_s = \rho c T_m$ and the minimum enthalpy for a purely melted substance to be $h_l = \rho c T_m + L_f$, we can assume that enthalpy is linearly distributed between these endmembers in partially molten medium

$$f = \begin{cases} 0 & \text{if } h < h_s, \\ \frac{h - h_s}{h_l - h_s} & \text{if } h_s \le h \le h_l \\ 1 & \text{otherwise,} \end{cases} \quad (15.102)$$

where $h = \rho c T + L_f f$. That should not be satisfying because the melt fraction is computed from the local enthalpy, which itself depends on the melt fraction. This circularity calls for an iterative solution procedure, where a guess in melt fraction change with time allows us to make a first estimate of the updated temperature at any given point, which we use then to correct for the enthalpy to recalculate the local melt fraction. The iterations between temperature and melt fraction are continued for as long as needed (i.e., until both are consistent with each other) before proceeding to a new time step. The implementation and algorithm are presented in Section 15.5.1.

The interesting aspect of this problem is that an analytical solution exists. Assuming, for example, that the pure substance is initially purely solid at its melting temperature T_m and the prescribed temperature at one of the boundaries is raised to $T_b > T_m$ yields a famous mathematical problem often referred to as the Stefan problem. Considering an infinite half-space with $T_b > T_m$ imposed at $x = 0$, an initial condition $T(x, t = 0) = T_m$ and $T_\infty = T_m$ at $x \to \infty$, the analytical solution to the temperature profile is

$$T(x,t) = \begin{cases} T_m & \text{if } f \leq 0 \\ T_b - (T_b - T_m)\frac{\text{erf}(x/2\sqrt{\kappa t})}{\text{erf}(\lambda)} & \text{otherwise.} \end{cases} \tag{15.103}$$

The position of the melting front follows

$$x_m(t) = 2\lambda\sqrt{\kappa t}, \tag{15.104}$$

where the constant λ is computed from the solution of the equation

$$\lambda \exp\left(\lambda^2\right) \text{erf}(\lambda) = \frac{St}{\sqrt{\pi}}, \tag{15.105}$$

and St is a dimensionless number called the Stefan number defined as $St = c(T_b - T_m)/L_f$.

15.5.1 Implementation

Solving simultaneously for the temperature and melt fraction introduces a degree of complexity in the design of the scheme. We proceed with an iterative approach, which means that at each new time step, a first estimate of the temperature at each grid point allows for a new estimate of the melt fraction to be used to compute the $\partial f/\partial t$ term of the PDE. In turn, this allows us to refine the estimate of the temperature, and so forth. This cycle of iterations occurs at a fixed time and we can proceed to the next time step once the solution to the melt fraction and temperature are consistent enough at every point in the domain. How does that work?

First, we write discretize the PDE

$$\frac{T_k^{n+1} - T_k^n}{\Delta t} = \frac{\kappa \Delta t}{\Delta x^2}\left(T_{k+1}^{n+1} - 2T_k^{n+1} + T_{k-1}^{n+1}\right) - \frac{L_f}{c\Delta t}\left(f_k^{n+1} - f_k^n\right), \tag{15.106}$$

which is an implicit scheme for T. We now introduce a second superscript index m where $0 \le m \le m_{max}$ is a counter for the number of iterations used for T and f to converge and yield a consistent result. The FDE becomes

$$\frac{T_k^{n+1,m+1} - T_k^n}{\Delta t} = \frac{\kappa \Delta t}{\Delta x^2}\left(T_{k+1}^{n+1,m+1} - 2T_k^{n+1,m+1} + T_{k-1}^{n+1,m+1}\right) - \frac{L_f}{c\Delta t}\left(f_k^{n+1,m} - f_k^n\right). \tag{15.107}$$

Note here that we do not specify the iteration counter $(m, m-1...)$ for the variables T and f taken from the previous time step n, as we consider those to be the values obtained after the iterations converged during the previous time step (i.e., they do not depend on the current count of iterations). When $m = 0$, as a guess for the first iteration of time step n we assume $f_k^{n+1,m} = f_k^n$ and then proceed to compute all the $T_k^{n+1,1}$ from that first guess. That step can be done with the TDMA as the tridiagonal matrix system is identical to that described in the implicit scheme. Once the $T_k^{n+1,1}$ is computed (i.e., satisfies the two BCs), the local enthalpy for the first iteration at time $n+1$ can be estimated

$$h_k^{n+1,1} = \rho c T_k^{n+1,1} + L_f f_k^n. \tag{15.108}$$

We can now compute an update to the local melt fraction

$$f_k^{n+1,1} = \begin{cases} 0 & \text{if } h_k^{n+1,1} < h_s, \\ \frac{h_k^{n+1,1} - h_s}{h_l - h_s} & \text{if } h_s \le h_k^{n+1,1} \le h_l \\ 1 & \text{otherwise.} \end{cases} \tag{15.109}$$

With this updated melt fraction we can improve our first estimate and compute $T_k^{n+1,2}$ $(m = 2)$ with (Eq. 15.107) where the latent heat term $\frac{L_f}{c\Delta t}\left(f_k^{n+1,m} - f_k^n\right)$ is likely no longer null at or next to the melting front. These internal iterations (all at time step $n+1$) are stopped when the relative local enthalpy updates becomes smaller than a tolerance

$$\left| \frac{h_k^{n+1,m} - h_k^{n+1,m-1}}{h_k^{n+1,m-1}} \right| < \text{Tol}, \tag{15.110}$$

at every position k. The pseudo-code that follows gives an idea for the design of the algorithm

```
Tol=1e-3; % tolerance
dt=...;
dx=...;
D=...;
Lf=...;
rho=...;
c=...;
Diff_n=D*dt/dx^2;
T_old=...; % Initial condition for T
h_old=...; % Initial condition for h
f_old=...; % initial condition for f
M=... % build matrix M or M'
M_BE = ... % inversion of M or M'
for time=1:maxtime-1 % Marching equation
        T_prev=T_old % Tprev is iteration m-1
        h_prev=h_old;
        f_prev=f_old;
        h_new=h_prev+2*Tol; % to start the next loop
        Source=0;
        while norm(h_new-h_prev./h_prev) >Nx*Tol
                RHS=T_old+V_BC+Source; % modify RHS with
                BC and Source
                T_new=M_BE *RHS; % tri-diagonal solver
                h_new=rho*c*T_new+Lf*f_prev;
                f_new-...
                Source=-Lf/c*(f_new-f_old);
                T_prev=T;
                f_prev=f_new;
                h_prev=h_new;
        end
        T_old=T_new;
        f_old=f_new;
        h_old=h_new;
end
```

15.5.2 Results

In this section, the Stefan problem (half-space melting) is solved in one dimension with an implicit diffusion scheme and an iterative method for the melt fraction–temperature coupled variables. The numerical results are compared to the analytical solution for the temperature profile (Eq. 15.103) and the position of the melting front with time (Eq. 15.104) in Figure 15.8. In this particular example, the tolerance for convergence is set to 10^{-3} for the relative convergence of the enthalpy, the diffusion number $\mathcal{D} = \kappa\Delta t/\Delta x^2 = 0.01$ and the Stefan number $St = 0.1$. The comparison with the analytical solution is excellent, but the conditions for convergence require a significant amount of iterations per time step (between 50 and 100). The convergence becomes easier with time progressing because the heat flux into the solid domain decreases, leading to less melting per time step.

15.6 Two-dimensional diffusion equation

Many applications of diffusion in Earth and planetary sciences require two or more dimensions. We focus exclusively on two-dimensional problems here. The governing equation is in general

$$\frac{\partial \rho}{\partial t} = \nabla \cdot (\mathbf{D}\nabla \rho). \tag{15.111}$$

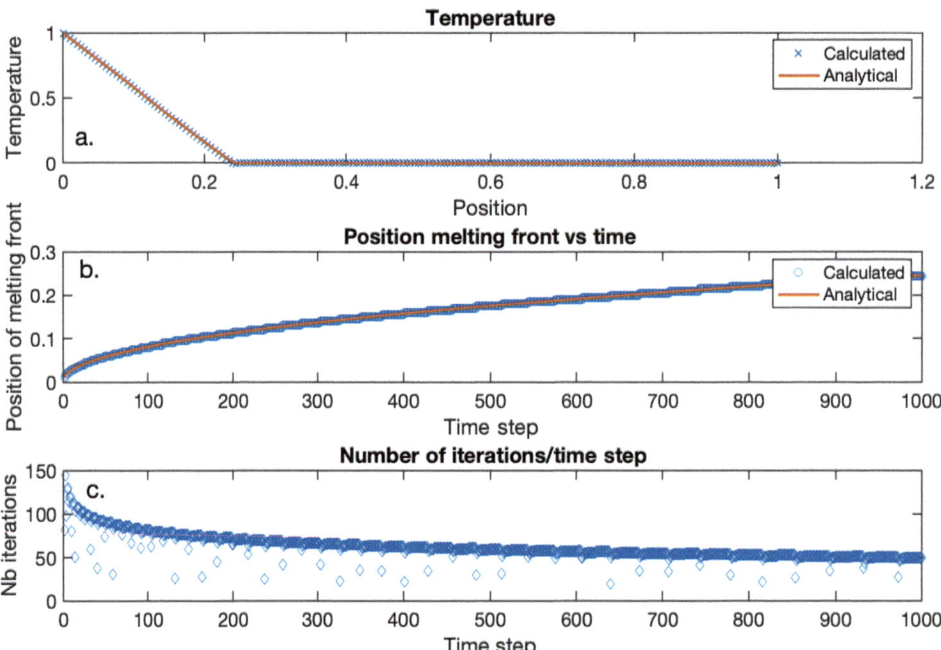

Figure 15.8 Solution to the Stefan problem using an implicit scheme for the diffusion equation and an iterative scheme for the partitioning between sensible and latent heat. The run uses 200 grid points and 1000 time steps. a): The temperature profile after a dimensionless time $Fo = Dt/L^2 = 0.003$, using a Stefan number $St = 0.1$. b): The position of the melt–solid interface over time. c): The number of iterations required for convergence at each time step.

In the two-dimensional case and focusing at this time on a Cartesian frame of reference, this equation becomes

$$\frac{\partial \rho}{\partial t} = \frac{\partial}{\partial x}\left(D_x \frac{\partial \rho}{\partial x}\right) + \frac{\partial}{\partial y}\left(D_y \frac{\partial \rho}{\partial y}\right). \tag{15.112}$$

For simplicity, we further assume that the diffusion coefficients D_x, D_y, are homogeneous in space, although it is important to note that the methods outlined next can easily be extended beyond this limitation.

In the previous section, when solving the one-dimensional diffusion equation we found that an implicit approach was worth the effort as it provided unconditional convergence. Here we strive to find a method that offers a similar advantage.

15.6.1 Alternative direction implicit (ADI) method

The main challenge in developing an implicit approach for a two-dimensional problem is that we lose the simple tridiagonal matrix structure that allowed for a very efficient solution. An implicit FDE here would lead to

$$\rho_{k,l}^{n+1} - \rho_k = \mathcal{D}_x\left(\rho_{k+1,l}^{n+1} - 2\rho_{k,l}^{n+1} + \rho_{k-1,l}^{n+1}\right) + \mathcal{D}_y\left(\rho_{k,l+1}^{n+1} - 2\rho_{k,l}^{n+1} + \rho_{k,l-1}^{n+1}\right), \tag{15.113}$$

where $\rho_{k,l}^n = \rho(x_k, y_l, t_n)$. We can recast this equation in a matrix–vector product format by defining a column vector with $Nx \times Ny$ rows

$$\rho^n = \begin{pmatrix} \rho_{1,1}^n \\ \rho_{2,1}^n \\ \vdots \\ \rho_{Nx,1}^n \\ \vdots \\ \rho_{1,2}^n \\ \vdots \\ \rho_{Nx,2}^n \\ \vdots \\ \vdots \\ \rho_{Nx,Ny}^n \end{pmatrix} \tag{15.114}$$

$$\mathbf{M}\rho^{n+1} = \rho^n, \tag{15.115}$$

where $\mathbf{M}$ is no longer tridiagonal as three secondary diagonal "bands" arise from the two components of the second derivative over space of ρ. This makes the tridiagonal inversion solver irrelevant for the inversion of the matrix $\mathbf{M}$. This can be done either with a brute force inversion or iteratively (e.g., the Jacobi method). In this section, we avoid this problem altogether and focus on an alternative approach, where we split the solution of the spatial derivative by conducting a first half-step of the time integration over a dimension implicitly and the other explicitly and then a second half-step where the choice of spatial dimension integrated implicitly versus explicitly is switched. This method is generally referred to as the alternative direction implicit (ADI) approach. We present the approach first, discuss the performance of the algorithm, and then present two applications: diffusion of trace elements in a two-dimensional crystal and a hillslope diffusion model of crater degradation.

The main idea behind ADI is to compute two half-steps that alternate the direction treated implicitly in the integration. For instance, integrating the diffusion equation from time n to $n+1$, we would first solve

$$\frac{T^*_{k,l} - T^n_{k,l}}{\Delta t/2} = \frac{D_x}{\Delta x^2}\left(T^*_{k+1,l} - 2T^*_{k,l} + T^*_{k-1,l}\right) + \frac{D_y}{\Delta y^2}\left(T^n_{k,l+1} - 2T^n_{k,l} + T^n_{k,l-1}\right), \tag{15.116}$$

where $T^*_{k,l} = T^{n+1/2}_{k,l}$. Then the second half-step considers the other direction for the implicit integration

$$\frac{T^{n+1}_{k,l} - T^*_{k,l}}{\Delta t/2} = \frac{D_x}{\Delta x^2}\left(T^*_{k+1,l} - 2T^*_{k,l} + T^*_{k-1,l}\right) + \frac{D_y}{\Delta y^2}\left(T^{n+1}_{k,l+1} - 2T^{n+1}_{k,l} + T^{n+1}_{k,l-1}\right). \tag{15.117}$$

Each of these half-steps can be solved with a tridiagonal solver similar to the one-dimensional diffusion equation. An important final adjustment with the ADI method is to avoid biases with solving implicitly over one spatial dimension first and the other as a second step. This is done by alternating the order of the half integration steps at each time step. In other words, if x is solved first implicitly at a time step, over the next time step we would select y to be solved first implicitly.

The implementation of the ADI scheme involves two tridiagonal matrices, here called $\mathcal{M}_x$ and $\mathcal{M}_y$ of dimension $Nx \times Nx$ and $Ny \times Ny$, respectively. Each of these matrices are similar to those we built earlier for one-dimensional diffusion with the implicit scheme.

$$\mathbf{M}_x = \begin{pmatrix} 1+2\mathcal{D}_x & -\mathcal{D}_x & 0 & \dots & & \\ -\mathcal{D}_x & 1+2\mathcal{D}_x & -\mathcal{D}_x & 0 & \dots & \\ 0 & -\mathcal{D}_x & 1+2\mathcal{D}_x & -\mathcal{D}_x & 0 & \dots \\ \vdots & & & & & \\ & & \dots & 0 & -\mathcal{D}_x & 1+2\mathcal{D}_x \end{pmatrix} \tag{15.118}$$

and

$$\mathbf{M}_y = \begin{pmatrix} 1+2\mathcal{D}_y & -\mathcal{D}_y & 0 & \dots & & \\ -\mathcal{D}_y & 1+2\mathcal{D}_y & -\mathcal{D}_y & 0 & \dots & \\ 0 & -\mathcal{D}_y & 1+2\mathcal{D}_y & -\mathcal{D}_y & 0 & \dots \\ \vdots & & & & & \\ & & \dots & 0 & -\mathcal{D}_y & 1+2\mathcal{D}_y \end{pmatrix}. \tag{15.119}$$

A pseudo-code illustrating the implementation of ADI is shown below.

```
test=0;
for time=1:maxtime-1 % Marching equation
        if test==0
                for l=1:ny
                        R1=rho(k,l)*(1-Dy)+Dy/2*(rho(k,l-1)
                        +rho (k,l+1));
                        rho_half=invM_x *R1;
                        % use tri-diagonal solver
                end
                for k=1:nx
                        R2=rho(k,l)*(1-Dx)+Dx/2*(rho(k-1,l)
                        +rho (k+1,l));
                        rho_new=invM_y *R2; % tri-diagonal solver
                end
                test=1;
        else
                for k=1:nx
                        R2=rho(k,l)*(1-Dx)+Dx/2*(rho(k-1,l)
                        +rho (k+1,l));
                        rho_half=invM_y *R2; % tri-diagonal solver
                end
                for l=1:ny
                        R1=rho(k,l)*(1-Dy)+Dy/2*(rho(k,l-1)
                        +rho (k,l+1));
                        rho_new=invM_x *R1; % tri-diagonal solver
                end
                test=0;
        end
        rho=rho_new;
end
```

It is easy to show by extension of the schemes presented for one-dimensional diffusion that the scheme proposed here is consistent. Moreover, although not shown here, the ADI method preserves the unconditional stability of an implicit scheme.

15.6.2 Diffusion of trace elements in minerals

Equilibrium conditions are often assumed when constraining the conditions of magma storage or metamorphism in collected samples. It allows for the use of thermodynamic principles to infer intensive variables (e.g., pressure and temperature conditions) given the assumption of chemical and thermal equilibrium among phases comprised in the sample analyzed. Nevertheless, signs of chemical disequilibrium are ubiquitous and, if kinetics adds some complexity in assessing the intensive variables that prevailed during the formation of the sample, it offers the opportunity to study rates and timescales, because a system out of equilibrium is bound to relax to equilibrium given enough time. For example, if, through careful laboratory experiments, we can constrain the factors controlling the rate of relaxation towards equilibrium (which depend on temperature, composition, and pressure), then any preserved (frozen) measure of disequilibrium could provide an estimate for the time the system has spent out of equilibrium under a given set of conditions. In essence, this is the leading principle behind diffusion chronometry, a field that has provided a wealth of information over the past decades. For more information, the interested reader is referred to seminal papers and recent reviews in the field [3, 5, 6, 10, 22, 27].

Relaxation of chemical heterogeneities within crystals occurs by diffusion and it is driven by local gradients in composition (more accurately, it is driven by gradients in chemical potential). The main idea behind diffusion chronometry is to measure compositional transects at high resolution and track for example trace element concentration along these transects. Under an assumed initial concentration of the same element along that transect, and assuming a value for the diffusion coefficient of that element, the duration required to reach the final/measured concentration transect starting from the initial condition provides a timescale estimate for the process of interest (time available for the mineral to re-equilibrate from the initial disequilibrium). The approach comes with a suite of assumptions, some easier to justify than others. It is commonly assumed, for simplicity, that diffusion and growth are decoupled (i.e., the process of diffusion is not affected by a moving boundary, unlike the Stefan problem). Another source of uncertainty is related to the fact that solving for diffusion requires one initial condition (concentration at time 0 within the crystal or zone of interest) and two BCs. The initial condition can be quite complex and hard to constrain when information is lacking, and a step function is generally assumed to provide an upper bound for the timescale of interest. The BCs are also challenging as the melt composition in contact with the crystal at the time of diffusion cannot be measured. A common and justifiable assumption is that the concentration at the boundary remains constant over time because the rate of diffusion in melts/liquids around crystals tends to me several orders of magnitude faster than in crystals.

Diffusion coefficients can depend on a large number of factors, including temperature (strongly), pressure, and composition. Isothermal conditions are often used to compute these timescales, with users providing rough bounds on the timescale estimates from simulated diffusion at lower temperature (longer timescale) and higher temperature (shorter timescale). Finally, the geometry of the crystal and the crystal zone over which diffusion is acting is often considered planar and distant from any corners or heterogeneity that may impact the diffusion process.

In this section we provide a simple example where diffusion within a rectangular (two-dimensional) crystal is considered. We assume a constant concentration outside the crystal ($C_b = 0$) and a homogeneous initial concentration ($C_0 = 1$) within the crystal. The simulations are isothermal, and we therefore assume that the diffusion coefficient does not change with time. Finally, we assume the diffusion coefficients to be homogeneous in space. We contrast two cases: isotropic diffusion ($D_x = D_y$) and anisotropic diffusion ($D_x = 6D_y$). The diffusion equation for both cases is

$$\frac{\partial C}{\partial t} = D_x \frac{\partial^2 C}{\partial x^2} + D_y \frac{\partial^2 C}{\partial y^2}. \tag{15.120}$$

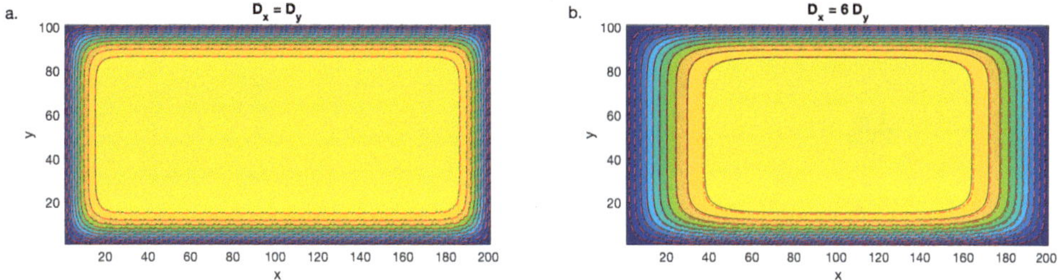

Figure 15.9 Comparison of the numerical solution to two-dimensional diffusion out of a rectangle for isotropic (a) and anisotropic (b) diffusion. The dashed lines show contours of constant concentration from the analytical solution.

The analytical solution is

$$C(x,y,t) = \frac{16C_0}{\pi^2}\sum_{l,m}\frac{(-1)^{l+m}}{(2l+1)(2m+1)}\cos\left(\frac{(2l+1)\pi x}{L_x}\right)\cos\left(\frac{(2m+1)\pi y}{L_y}\right)$$

$$\exp\left(-\frac{\pi^2(2l+1)^2D_xt}{L_x^2}\right)\exp\left(-\frac{\pi^2(2m+1)^2D_yt}{L_y^2}\right), \tag{15.121}$$

where L_x and L_y are the dimensions of the crystal along the x and y axes respectively. Figure 15.9 shows comparisons between the numerical and analytical solutions. The latter is plotted as red isoconcentration dashed lines on top of the numerical solution. Panel (a) shows the case with isotropic diffusion and panel (b) with $D_x = 6D_y$.

15.6.3 **Hillslope diffusion (linear)**

The evolution of soil-mantled hilltops is a component of soil mass transport across a landscape. It has received a large amount of attention for over a century as landscape evolution is tied to many fundamental processes in Earth sciences, soil production (mechanical and chemical erosion), regional tectonic processes (e.g., uplift), sediment transport budgets, and biomass activity. Davis and Gilbert [8; 12] first used the topographic convexity of hillslopes near hilltops to qualitatively infer that soil creep was playing an important role in mobilizing soils. If the creep rate (in terms of mass flux of soil) is made proportional to the slope

$$\mathbf{q} = -\Omega\underbrace{\nabla h}_{slope}, \tag{15.122}$$

where h is the elevation, q is the mass flux in terms of *mass/distance/time*, and Ω is the constant of proportionality. Given that ∇h is dimensionless, Ω has the same units as q, it can therefore be decomposed into

$$\Omega = \rho D, \tag{15.123}$$

where ρ is the soil density and D a diffusion coefficient. The elevation is a function of the position (x, y), therefore

$$\nabla h = \begin{pmatrix}\frac{\partial h}{\partial x}\\ \frac{\partial h}{\partial y}\end{pmatrix}. \tag{15.124}$$

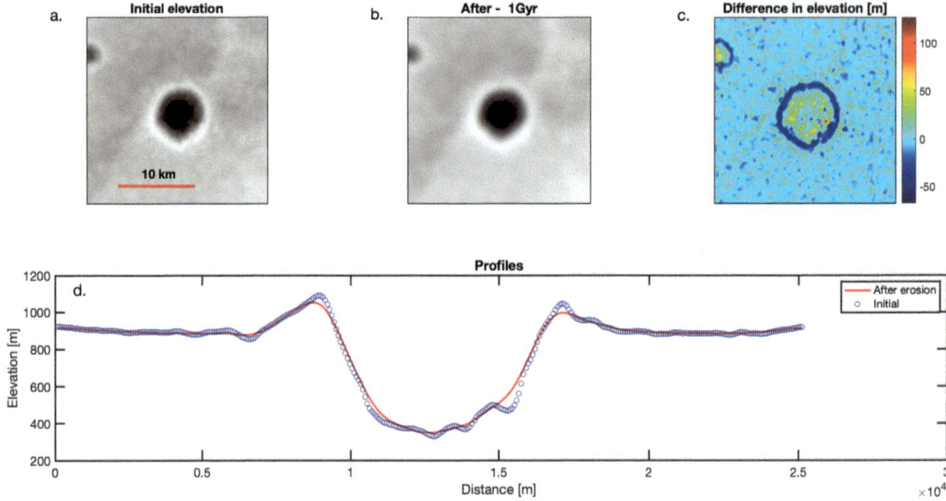

Figure 15.10 Modeled diffusion by micrometeorite impacts on a crater located in Tyrrhena Terra on Mars (topography obtained with High Resolution Stereo Camera (HRSC) DEM). Diffusion erases first topographic features with short wavelength. The author acknowledges the help of Dr. Ben Boatwright for the selection and pre-processing of the image.

The mass of soil in a small soil column of thickness h and areal dimension $\Delta x \times \Delta y$ is

$$M(x, y, t) = \rho h(x, y, t)\Delta x \Delta y, \tag{15.125}$$

where the density ρ is assumed constant and $h(x, y, t)$ is to be understood as the average elevation of the soil column in small control surface area $\Delta x \times \Delta y$ centered at (x, y). The mass budget of soil transported through that column during an interval of time Δt (similar to our derivation of the general diffusion equation) is

$$\begin{aligned} &\rho(h(x, y, t+\Delta t) - h(x, y, t))\Delta x \Delta y = \qquad (15.126)\\ &-\Delta t\big(\Delta y(q_x(x+\Delta x/2, y, t) - q_x(x-\Delta x/2, y, t)) + \Delta x(q_y(x, y+\Delta y/2, t) - q_y(x, y-\Delta y/2, t))\big). \end{aligned}$$

Dividing both ends with $\Delta t \times \Delta x \times \Delta y$ and taking the limit when each become vanishingly small, we retrieve the familiar diffusion equation

$$\frac{\partial h}{\partial t} = D\left(\frac{\partial^2 h}{\partial x^2} + \frac{\partial^2 h}{\partial y^2}\right). \tag{15.127}$$

For a more detailed discussion of the diffusion equation applied to landscape evolution, the interested reader is directed to Pelletier [18]. An important caveat of the diffusion model with respect to soil transport dynamics is that it does fail when dealing with steeper slopes where the diffusive behavior of mass transport is no longer a good assumption. Readers are directed to the following papers for more detailed discussions of more advanced (nonlinear) mass transport models [11, 13, 19, 21].

For this simple application of the two-dimensional diffusion equation and the ADI method, we consider erosion of old craters on planetary surfaces (here Mars) by micro-meteorite impacts. The model we propose uses a High Resolution Stereo Camera (HRSC) DEM with a resolution of 75 m/pix (provided kindly by Dr. Ben Boatwright) from Tyrrhena Terra on Mars. The size of the image is about 25km × 25km. The boundary conditions on each of the four edges ($x = 0$, $x = Nx + 1$, $y = 0$, and $y = Ny + 1$) is set to a no-flux boundary (no mass loss by the domain). The diffusion coefficient is homogeneous and isotropic ($D_x = D_y = D$) in the calculations and was set to $7 \times 10^{-5} m^2$/yr, an arbitrary value (perhaps on the high side for diffusion induced by micro-meteoric impacts). Not surprisingly, the simulation shows the effect of diffusion erasing topographic features at a rate that is controlled by the wavelength λ of the features (damping is proportional to λ^{-2};, see section about numerical diffusion at the beginning of Part III of the textbook).

Chapter 16

One-dimensional advection–diffusion equation

16.1 Derivation from material derivatives

In continuum mechanics, it is sometimes advantageous to use a reference frame coupled to the motion of the material we study rather than fixed in space (say in the lab where the experience is performed). This introduces the concept of Eulerian (e.g., fixed in the lab) versus Lagrangian (coupled to the flow of matter) reference frames. A good example of the difference between these frames of reference is to consider the flow field at the surface of a river and comparing the view points of an observer on the shore and someone flowing down on a raft. For the latter, the observer on the shore is moving upstream (Lagrangian viewpoint), while for the observer the person on the raft is following the water downstream. The physical problem is equivalent, but the mathematical description differs. Let us define a scalar field of interest that can vary in time and space, such as the water temperature or the concentration of a dissolved contaminant. We define the material derivative of the concentration of contaminant $T(x, t)$ as the time derivative of the field T that is measured by the person on the raft (Lagrangian reference frame). It is symbolized as DT/Dt. We consider all the matter within an arbitrary volume of water V around the raft that we will track over time, that is, as if we tagged each molecule and keep track of its progress. The mass of contaminant within that volume at time t is

$$M(t) = \int_{V(t)} T(x, t)\rho \mathrm{dV}, \tag{16.1}$$

where T is a concentration (i.e., mass of contaminant/mass of water) and ρ is the density of the water (here assumed constant over time and space for simplicity). After an interval of time Δt the volume of water considered moves and possibly deforms $V(t) \rightarrow V(t + \Delta t)$, the mass of contaminant within that parcel of water changes to

$$M(t + \Delta t) = \int_{V(t+\Delta t)} T(x, t + \Delta t)\rho \mathrm{dV}. \tag{16.2}$$

Note that, unlike the advection equation derived earlier, the mass of contaminant is not necessarily conserved in our test volume moving with the water

$$M(t) \neq M(t + \Delta t), \quad \Delta M = M(t + \Delta t) - M(t) \neq 0. \tag{16.3}$$

There exist two possible reasons why $\Delta M \neq 0$. First, there is a source or sink of contaminant within our arbitrary deforming volume that adds or subtracts mass (we discard this eventuality for the moment). Second, contaminant can diffuse through the boundary of the defined volume of water

Introduction to Numerical Modeling in Earth and Planetary Sciences, Christian Huber, Oxford University Press.
© Christian Huber (2025). DOI: 10.1093/oso/9780198802716.003.0017

that we take as a reference. It is important to realize that contaminant gain/loss by advection (i.e., flow of water) is not possible because the boundary of our integration volume are deformed with the flow of water. In other words, the boundaries of our control volume move at the same pace as the water locally. Taking the contaminant diffusion flux as the source of the potential imbalance, we get

$$\Delta M = -\Delta t \int_{\Omega} \mathbf{q} \cdot \mathbf{n} \mathrm{d}\Omega, \tag{16.4}$$

where Ω denotes the boundary of the deforming volume V, $\mathbf{n}$ is the outward pointing normal to the boundary Ω, and $\mathbf{q}$ is the diffusive flux at the boundary. Using Fick's first law

$$\mathbf{q} = -\mathcal{D}\nabla T, \tag{16.5}$$

where $\mathcal{D}$ is a diffusion coefficient. Dividing by Δt, we obtain

$$\frac{\Delta M}{\Delta t} = -\int_{\Omega} \mathbf{q} \cdot \mathbf{n} \mathrm{d}\Omega, \tag{16.6}$$

which, if we take the limit $\Delta t \rightarrow 0$ and use the divergence theorem (see Part I for an overview of vector calculus), yields an expression that introduces the material derivative of the mass M

$$\frac{DM}{Dt} = -\int_{V} \nabla \cdot \mathbf{q} \,\mathrm{dV}. \tag{16.7}$$

If water is assumed incompressible (we assumed this earlier), then the body of water considered moves downstream and deforms but contains the same volume of water $V(t + \Delta t) = V(t)$ and we get

$$\int_{V} \left[\frac{DT}{Dt} + \nabla \cdot \mathbf{q} \right] \mathrm{dV} = 0, \tag{16.8}$$

which is true for any arbitrary control volume V, which implies that the integral's argument is identically zero and

$$\frac{DT}{Dt} = -\nabla \cdot \mathbf{q} = \nabla \cdot (\mathcal{D}\nabla T). \tag{16.9}$$

Assuming a homogeneous diffusion coefficient $\mathcal{D}$ and using Eq. 14.6, and considering a one-dimensional problem, we finally retrieve the one-dimensional advection–diffusion equation

$$\frac{\partial T}{\partial t} + u\frac{\partial T}{\partial x} = \mathcal{D}\frac{\partial T}{\partial x^2}. \tag{16.10}$$

16.2 **Examples of application of the advection–diffusion equation**

The advection–diffusion equation is ubiquitous as it pertains to the description of any conserved field where transport can be accommodated by either diffusion or advection. Sensible heat, that is, the energy stored by the heat capacity of a material, in a deforming medium follows this equation

$$\frac{\partial f}{\partial t} + u\frac{\partial f}{\partial x} = k\frac{\partial^2 f}{\partial x^2}, \tag{16.11}$$

where f is defined as the product of the material's density ρ, its specific heat c, and temperature T, $f = \rho * c * T$ and k is in this case assumed to be a homogeneous thermal conductivity. As we see in our derivation, the mass balance of chemical species in rivers, oceans, the atmosphere, or groundwater would also obey to the same overall conservation law. It is therefore important to identify the conditions under which different discretization schemes for this equation offer proper convergence. However, establishing convergence of the schemes is difficult mathematically.

16.3 Scheme 1: FE(time)–CE(space)

The discretization of the diffusion equation naturally leads to a centered scheme for the spatial part of the discretization, while a centered scheme for the advection equation was a source of problem. Given that the advection–diffusion equation is a blend of both equations, it is interesting to start our analysis with the centered scheme in space and forward (explicit) in time. The finite difference equation (FDE) for this scheme is

$$\frac{T_k^{n+1} - T_k^n}{\Delta t} + u\frac{T_{k+1}^n - T_{k-1}^n}{2\Delta x} = D\frac{T_{k+1}^n - 2T_k^n + T_{k-1}^n}{\Delta x^2}. \tag{16.12}$$

Multiplying by Δt on both sides and reorganizing some terms we get a marching equation of the form

$$T_k^{n+1} = T_k^n - \underbrace{\frac{u\Delta t}{2\Delta x}}_{C/2}\left(T_{k+1}^n - T_{k-1}^n\right) + \underbrace{\frac{D\Delta t}{\Delta x^2}}_{\mathcal{D}}\left(T_{k+1}^n - 2T_k^n + T_{k-1}^n\right), \tag{16.13}$$

where two old acquaintances are appearing, unsurprisingly: the Courant C and diffusion $\mathcal{D}$ numbers, which again shows that this equation is a blend of the advection and the diffusion equations. It is therefore expected that convergence may be dependent on the value of these two dimensionless parameters. Figure 16.1 shows the scheme, which is identical to that of the explicit diffusion scheme.

16.3.1 Performance

The model setup is a periodic one-dimensional domain with a Gaussian distribution $T(x, t = 0) = \exp(-x^2/(2\sigma))$ for the initial condition. The velocity field u is homogeneous and we fix the Courant number $C = 0.05$ in these calculations. We define a grid Peclet number $Pe_g = C/\mathcal{D}$ which we vary (i.e., we vary the diffusion number value $\mathcal{D}$). Figure 16.2 shows snapshots in time for two different values of the grid Peclet number (0.5 and 10). In both panels, the symbols show the solution of the FDE and are compared to solid lines showing the analytical solution

$$T(x, t) = \frac{\sqrt{\pi 2\sigma}}{\sqrt{\pi(2\sigma + 4Dt)}}\exp\left(-\frac{(x - ut)^2}{2\sigma + 4Dt}\right). \tag{16.14}$$

The finite difference solutions match the analytical solution well for $Pe_g = 0.5$ at all times, but the solution for $Pe_g = 10$ shows significant amounts of numerical dispersion (e.g., negative values

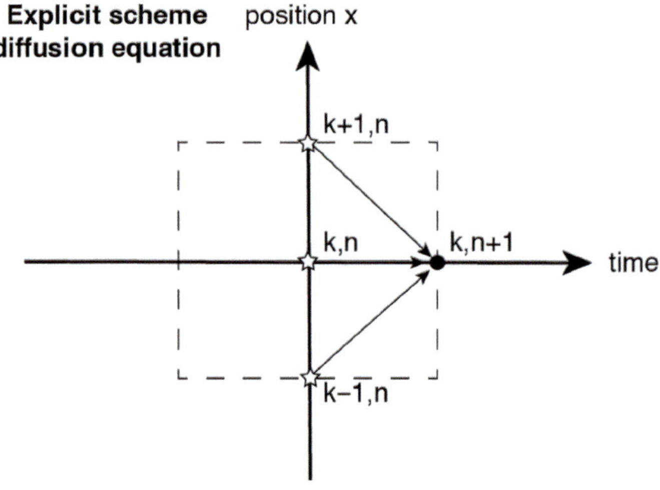

Figure 16.1 Stencil for the forward (time) and centered (space) advection–diffusion finite difference equation (FDE).

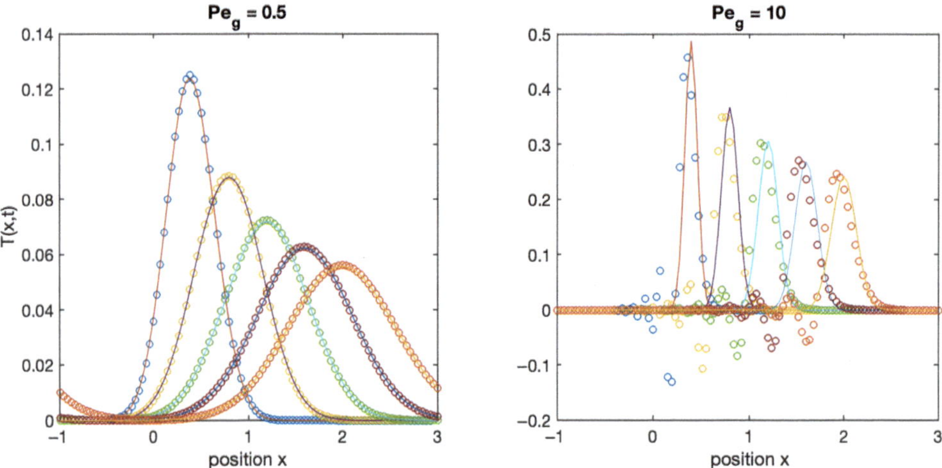

Figure 16.2 Outputs using the forward Euler (time) and centered Euler (space) finite difference scheme for the advection–diffusion equation with the Courant number $C = 0.05$ for two different values of the grid Peclet number Pe_g (0.5 on the left and 10 on the right). The symbols represent the solution of the finite difference equation (FDE) and the solid lines show the analytical solution.

of T). Both solutions are numerically stable. Specifically, the oscillations caused by numerical dispersion for $Pe_g = 10$ are damped with time and the solution remains bounded, even if not satisfying. As Pe_g is decreased below 0.1 or beyond 40, the scheme becomes unstable and the solutions diverge.

16.3.2 Consistency

The FDE involves several approximations of the field T at different times ($t = n\Delta t$ and $t = (n+1)\Delta t$. Using Taylor expansions we get

$$T_k^{n+1} = T_k^n + \frac{\partial T}{\partial t}\Delta t + \frac{\Delta t^2}{2}\frac{\partial^2 T}{\partial t^2} + \mathcal{O}(\Delta t^3) \tag{16.15}$$

$$T_{k+1}^{n} = T_k^n + \frac{\partial T}{\partial x}\Delta x + \frac{\Delta x^2}{2}\frac{\partial^2 T}{\partial x^2} + \mathcal{O}(\Delta x^3) \tag{16.16}$$

$$T_{k-1}^{n} = T_k^n - \frac{\partial T}{\partial x}\Delta x + \frac{\Delta x^2}{2}\frac{\partial^2 T}{\partial x^2} + \mathcal{O}(\Delta x^3). \tag{16.17}$$

The time component of the truncation error is

$$\mathcal{T}_t = \frac{\Delta t}{2}\frac{\partial^2 T}{\partial t^2} + \mathcal{O}(\Delta t^2). \tag{16.18}$$

The spatial contribution to the truncation error requires more work. The contribution from the advection term $u\frac{\partial T}{\partial x}$ is

$$u\frac{T_{k+1}^n - T_{k-1}^n}{2\Delta x} = u\frac{\partial T}{\partial x} + u\underbrace{\frac{\Delta x^2}{6}\frac{\partial^3 T}{\partial x^3}}_{\textit{numerical dispersion}} + \mathcal{O}(\Delta x^4). \tag{16.19}$$

Similarly, we can compute the contribution of the diffusion term

$$\frac{T_{k+1}^n - 2T_k^n + T_{k-1}^n}{\Delta x^2} = \frac{\partial^2 T}{\partial x^2} + \frac{\Delta x^2}{12}\frac{\partial^4 T}{\partial x^4} + \mathcal{O}(\Delta x^4). \tag{16.20}$$

The truncation error of the scheme is finally

$$\mathcal{T} = \frac{\Delta t}{2}\frac{\partial^2 T}{\partial t^2} + \mathcal{O}(\Delta t^2) - u\frac{\Delta x^2}{6}\frac{\partial^3 T}{\partial x^3} - \frac{\Delta x^2}{12}\frac{\partial^4 T}{\partial x^4} + \mathcal{O}(\Delta x^4). \tag{16.21}$$

The scheme is consistent, and it is easy to show that in the limit $\Delta t, \Delta x \to 0$ the truncation error $\mathcal{T} \to 0$. The truncation error is first order in time and second order in space.

16.3.3 Stability

The approach to establishing the conditions for stability are similar to the work we laid out for the diffusion equation. We will resort to von Neumann stability analysis introducing

$$T_k^n = \sum_j a_j^n \exp(i\pi j\Delta x k). \tag{16.22}$$

With this, we can estimate

$$T_k^{n+1} - T_k^n = \sum_j \left(a_j^{n+1} - a_j^n\right)\exp(i\pi j\Delta x k). \tag{16.23}$$

From the advection term we get

$$T_{k+1}^n - T_{k-1}^n = \sum_j a_j^n \exp(i\pi j\Delta xk)\left(\exp(i\pi j\Delta x) - \exp(-i\pi j\Delta x)\right). \quad (16.24)$$

Using now the identity

$$\exp(i\pi j\Delta x) - \exp(-i\pi j\Delta x) = 2i\sin(\pi j\Delta x), \quad (16.25)$$

we have

$$T_{k+1}^n = T_{k-1}^n = 2i\sum_j a_j^n \exp(i\pi j\Delta xk)\sin(\pi j\Delta x). \quad (16.26)$$

Repeating the same calculations that we did for the diffusion scheme, we get

$$T_{k+1}^n - 2T_k^n + T_{k-1}^n = \sum_j a_j^n \exp(i\pi j\Delta xk)\left(\exp(i\pi j\Delta x) - 2 + \exp(-i\pi j\Delta x)\right), \quad (16.27)$$

where we can use the trigonometric property

$$2\cos(i\pi j\Delta x) = \exp(i\pi j\Delta x) + \exp(-i\pi j\Delta x), \quad (16.28)$$

and define

$$\sigma_j = 2\left(1 - \cos(i\pi j\Delta x)\right), \quad (16.29)$$

to get

$$T_{k+1}^n - 2T_k^n + T_{k-1}^n = -\sum_j \sigma_j a_j^n \exp(i\pi j\Delta xk). \quad (16.30)$$

Gathering all these terms back into the FDE gives us

$$\sum_j \left(a_j^{n+1} - a_j^n\right)\exp(i\pi j\Delta xk) = -i\underbrace{\frac{u\Delta t}{\Delta x}}_{=C}\sum_j a_j^n \exp(i\pi j\Delta xk)\sin(\pi j\Delta x) - \underbrace{\frac{D\Delta t}{\Delta x^2}}_{=\mathcal{D}}\sum_j \sigma_j a_j^n \exp(i\pi j\Delta xk). \quad (16.31)$$

Using the same approach as for the diffusion equation, we now multiply each term with $\exp(i\pi m\Delta xk)$, use the orthogonality of these basis functions to isolate a single wavenumber m out of the sum, and get

$$a_m^{n+1} - a_m^n = -iCa_m^n\sin(\pi m\Delta x) - \mathcal{D}a_m^n\sigma_m. \quad (16.32)$$

For simple notation, we introduce $\phi_m = m\pi\Delta x$ and recast the previous equation, using the definition of σ_m as

$$a_m^{n+1} = a_m^n\left[1 - iC\sin\phi_m + 2\mathcal{D}(\cos\phi_m - 1)\right], \quad (16.33)$$

which shows that between consecutive time steps the amplitude of the Fourier series mode m changes by a complex factor that includes both Courant number C and diffusion number $\mathcal{D}$.

Stability would imply

$$\left|\frac{a_m^{n+1}}{a_m^n}\right| \leq 1 \quad \text{for all } m. \tag{16.34}$$

Given the ratio of amplitudes is complex (has a real and imaginary part), it is best represented as a unit circle (radius = 1) in the complex plane. The condition then

$$|1 - iC \sin \phi_m + 2\mathcal{D}(\cos \phi_m - 1)| \leq 1 \tag{16.35}$$

means that the left-hand side remains inscribed in the unit circle in the complex plane to ensure stability. Actually, the left-hand side describes the equation of an ellipse in the complex plane.

The ellipse intersects the real axis at +1 when $\phi_m = 0$, reaches a maximum imaginary value of $\pm C$ at $\phi_m = \pi/2$, and intersects the real axis again at a value $1 - 4\mathcal{D}$ (see Figure 16.3). As the ellipse and the unit circle meet at $\phi_m = 0$, we can guarantee that the ellipse is fully inscribed into the unit circle if: 1) its radius of curvature at $\phi_m = 0$ is smaller than 1, and 2) if $C \leq 1$. Both conditions are required. Using the definition of the radius of curvature of an ellipse it is possible to show that the first condition (when $C > 0$) reduces to

$$C \leq 2 \underbrace{\frac{\mathcal{D}}{C}}_{=1/Pe_g} \leq \frac{1}{C}. \tag{16.36}$$

Going back to our results in Figure 16.2, where $C = 0.05$, we find stability bounds $0.1 \leq Pe_g \leq 40$. Moreover, we see that numerical dispersion is expected since the advective term in the FDE introduces a truncation term $\partial^3 T/\partial x^3$.

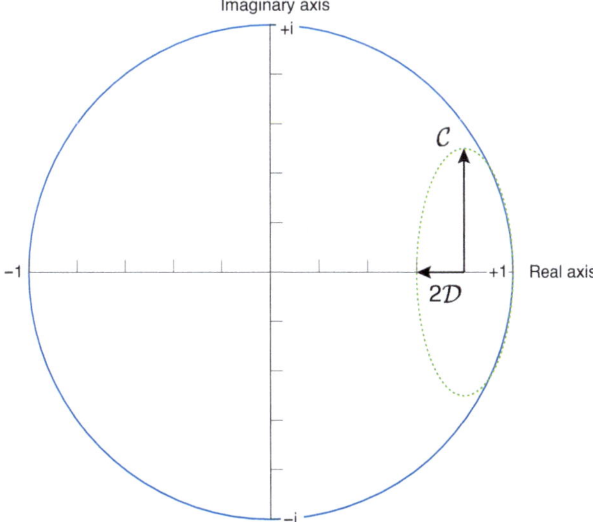

Figure 16.3 Diagram illustrating the condition for stability for the forward Euler (time)–centered Euler (space) advection–diffusion scheme. The unit circle in the complex plane is the space of admissible (stable) solutions and the ellipse shows the stability condition for the scheme with $C = 0.5$ and $\mathcal{D} = 0.1$. For these choices, the scheme is stable (conditional stability).

16.4 Scheme 2: FE(time)-upwind(space)

The upwind explicit (forward in time) scheme for the advection–diffusion equation leads to the FDE (here defined for a velocity $u > 0$)

$$\frac{T_k^{n+1} - T_k^n}{\Delta t} + u\frac{T_k^n - T_{k-1}^n}{\Delta x} = \mathcal{D}\frac{T_{k+1}^n - 2T_k^n + T_{k-1}^n}{\Delta x^2}. \tag{16.37}$$

In this section, we discuss the main difference with the centered scheme in terms of solution and stability. We see that by using the same trick as for the advection equation, we can recover the stability condition for the upwind scheme from our hard work on the centered scheme with little additional effort. Consistency is easy to show, the main difference being that the upwind scheme does not benefit from a second-order error in space, but is limited to first order in both time and space. However, one advantage of the upwind scheme is that the numerical dispersion that was plaguing the centered scheme is significantly reduced, mostly because the leading term of the spatial truncation error is a diffusive (damping term) as we see now. Taking the upwind FDE and recasting the advection term to

$$T_k^n - T_{k-1}^n = \frac{T_{k+1}^n - T_{k-1}^n}{2} - \frac{T_{k+1}^n - T_k^n + T_{k-1}^n}{2}, \tag{16.38}$$

we can now rewrite Eq. 16.37

$$T_k^{n+1} - T_k^n = -\frac{C}{2}\left(T_{k+1}^n - T_{k-1}^n\right) + \underbrace{\left(\mathcal{D} + \frac{C}{2}\right)}_{\mathcal{D}'}\left(T_{k+1}^n - 2T_k^n + T_{k-1}^n\right). \tag{16.39}$$

This finite difference expression is essentially identical to that of the centered scheme (Eq. 16.13), with $\mathcal{D} \to \mathcal{D} + C/2$. It shows that the upwind scheme introduces a spurious numerical diffusion equal to half of the Courant number, which increases damping of the solution. Unlike the advection equation, there is no way, with a finite velocity, to go around this numerical diffusion issue. Figure 16.4 provides an illustration of outputs from a set of calculations similar to the centered scheme. An obvious outcome is that the scheme here is seen to diffuse more than what is expected from the analytical expression. This spurious numerical diffusion depends directly on the choice of Courant number.

From the finite difference expression (Eq. 16.39), we can make the same stability argument as for the centered scheme, but with $\mathcal{D}'$ in the definition of the grid Peclet number instead of $\mathcal{D}$.

16.5 Implicit scheme

For the diffusion equation we saw that implicit schemes provided unconditional stability, at the expense of a slightly more complicated solution procedure (involving the inversion of a tridiagonal matrix system). It is interesting to pursue the same idea with the advection–diffusion equation. Skipping a few steps and writing the FDE for a centered in space implicit time scheme leads to

$$T_k^{n+1} - T_k^n = -\frac{C}{2}\left(T_{k+1}^{n+1} - T_{k-1}^{n+1}\right) + \mathcal{D}\left(T_{k+1}^{n+1} - 2T_k^{n+1} + T_{k-1}^{n+1}\right). \tag{16.40}$$

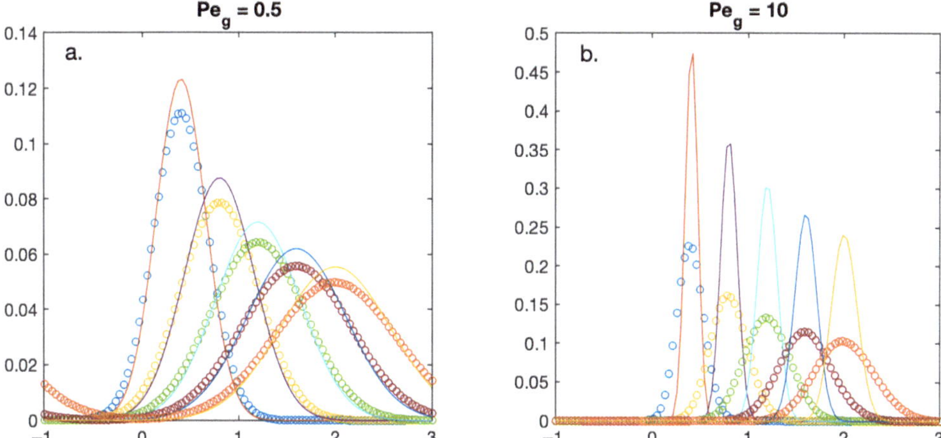

Figure 16.4 Outputs using the forward Euler (time) and upwind (space) finite difference scheme for the advection–diffusion equation with the Courant number $C = 0.05$ for two different values of the grid Peclet number Pe_g (0.5 on the left and 10 on the right). The symbols represent the solution of the finite difference equation and the solid lines show the analytical solutions at different times.

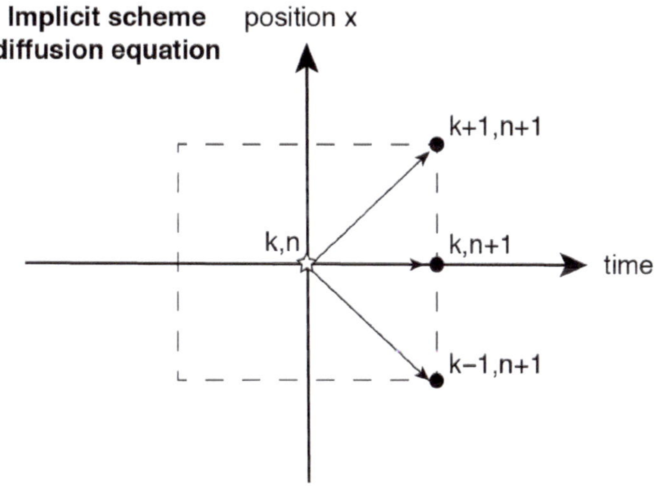

Figure 16.5 Stencil for the implicit (time) and centered (space) advection–diffusion finite difference equation (FDE).

The FDE for the implicit advection-diffusion scheme is identical to that for the implicit diffusion scheme (see Figure 16.5).

Using the same Fourier mode analysis for stability as for the explicit scheme with

$$T_k^n = \sum_j a_j^n \exp(i\pi j\Delta xk), \tag{16.41}$$

we retrieve now

$$a_m^{n+1} - a_m^n = -iC\sin(\pi\Delta xm)a_m^{n+1} - \mathcal{D}\sigma_m a_m^{n+1}. \tag{16.42}$$

We see that relative growth of the mode is complex and the stability condition becomes

$$\left|\frac{a_m^{n+1}}{a_m^n}\right| = \left|\frac{1}{1 + iC\sin\phi_m + \mathcal{D}\sigma_m}\right| \leq 1. \tag{16.43}$$

ϕ_m is defined again as $\pi m \Delta x$. This condition is satisfied if the norm of the denominator is > 1. The norm of a complex number is the square root of the product of the number and its complex conjugate; thus it follows that stability requires that

$$\sqrt{(1 + \mathcal{D}\sigma_m)^2 + C^2 \sin^2\phi_m} \geq 1. \tag{16.44}$$

This condition is always satisfied because $\mathcal{D}\sigma_m > 0$ and $C^2\sin^2\phi_m > 0$. The implicit centered scheme is therefore unconditionally stable for the advection–diffusion equation. Interestingly, the implementation of the scheme is very similar to the implicit scheme applied to the diffusion equation because the advection term only adds terms to the lower and upper diagonal of the matrix that needs to be inverted. It preserves the tridiagonal characteristic of the matrix. Reorganizing Eq. 16.40 with unknowns (time $n + 1$) on the left, we get

$$\left(\frac{C}{2} - \mathcal{D}\right) T_{k+1}^{n+1} + (1 + 2\mathcal{D})\, T_k^{n+1} + \left(-\frac{C}{2} - \mathcal{D}\right) T_{k-1}^{n+1} = T_k^n. \tag{16.45}$$

Ordering the Ts in terms of a column vector spanning all the positions within the domain ($k = 1, 2, \ldots Nx$)

$$\mathbf{T}^n = \begin{pmatrix} T_1^n \\ T_2^n \\ \vdots \\ T_{Nx}^n \end{pmatrix} \tag{16.46}$$

the marching equation for the implicit advection diffusion (centered) model becomes

$$\mathbf{M}\mathbf{T}^{n+1} = \mathbf{T}^n, \tag{16.47}$$

with a matrix $\mathbf{M}$ with dimensions $Nx \times Nx$

$$\mathbf{M} = \begin{pmatrix} 1+2\mathcal{D} & -\frac{C}{2}-\mathcal{D} & 0 & \ldots & & \\ \frac{C}{2}-\mathcal{D} & 1+2\mathcal{D} & -\frac{C}{2}-\mathcal{D} & 0 & \ldots & \\ 0 & \frac{C}{2}-\mathcal{D} & 1+2\mathcal{D} & -\frac{C}{2}-\mathcal{D} & 0 & \ldots \\ \vdots & & & & & \\ & & \ldots & 0 & \frac{C}{2}-\mathcal{D} & 1+2\mathcal{D} \\ & & & & & \end{pmatrix}. \tag{16.48}$$

$\mathbf{M}$ is a tridiagonal matrix. It can be inverted efficiently with the Thomas algorithm (see Section 15.3.2). The marching equation at each time step for the implicit scheme is therefore

$$\mathbf{T}^{n+1} = \underbrace{\mathbf{M}^{-1}}_{\equiv \mathbf{M}_{BE}} \mathbf{T}^n. \tag{16.49}$$

An illustration of the implicit scheme performance is shown in Figure 16.6. The spurious numerical diffusion is limited compared to the upwind scheme, although it retains the numerical dispersion observed in the explicit version of the scheme. This is not surprising because the spatial component of the truncation error is mostly identical to the explicit scheme, with a leading term $\partial^3 T/\partial x^3$.

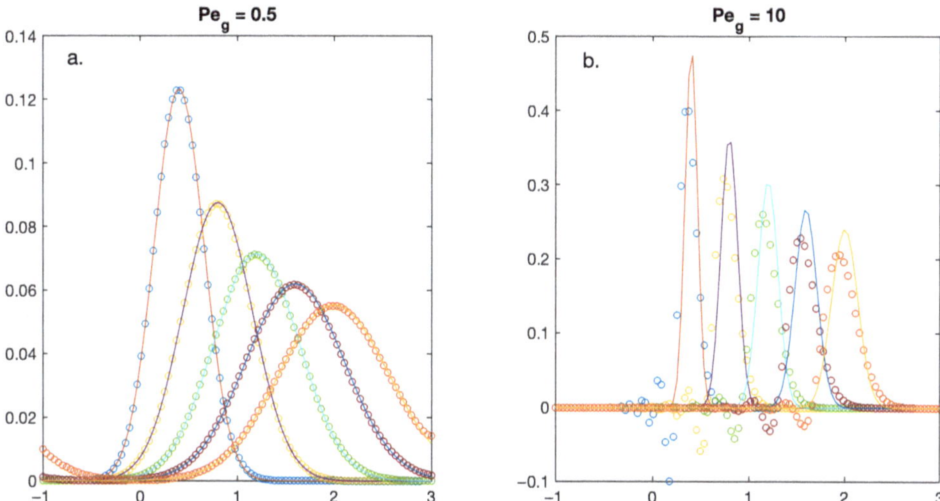

Figure 16.6 Outputs using the implicit (time) and centered (space) finite difference scheme for the advection–diffusion equation with the Courant number $C = 0.05$ for two different values of the grid Peclet number Pe_g (0.5 on the left and 10 on the right). The symbols represent the solution to the finite difference equation (FDE) and the lines are analytical solutions at different times.

16.6 **Exercises**

1. Breakthrough curves represent a time series of dissolved chemical species passing through a permeable porous medium. Use an advection–diffusion one-dimensional model to compute the breakthrough curves for a Gaussian pulse of contaminant at the inlet of the medium. The discharge through the medium provides the fixed and constant (homogeneous) velocity u in the advection–diffusion equation. Assume the inlet to be an imposed contaminant concentration with time (Gaussian) and that the flux of contaminant $q = uC - D\partial C/\partial x$ satisfies $\partial q/\partial x = 0$ at the outlet. Fixing the length of the domain to $L = 1$ m and playing with the Peclet number defined here as $Pe = uL/D$, where D is the diffusion/dispersion coefficient for the contaminant in the porous medium, measure the maximum amplitude of contaminant concentration at the outlet and timing of the peak concentration for various values of Pe. Discuss your results.
2. For a set of diffusion coefficient and advective velocity, show mathematically and, if possible numerically, that when the lengthscale over which transport is occurring increases, diffusion transport provides a diminishing contribution to the overall mass transport. How does the relative contribution of diffusive to advective transport vary with transport length?

Chapter 17

One-dimensional wave equation

17.1 Introduction to the wave equation

We focus here on a one-dimensional version of the wave equation, initially with constant/homogeneous material properties, an assumption we relax below. We start with the momentum conservation equation, here stated as Newton's second law

$$\underbrace{\frac{\partial \rho v_x}{\partial t}}_{inertia} = \frac{\partial \overbrace{\sigma_{xx}}^{\text{elastic stress}}}{\partial x}, \tag{17.1}$$

where v_x is the velocity

$$v_x = \frac{\partial u_x}{\partial t}, \tag{17.2}$$

with u_x is the displacement along the x-direction. The stress component

$$\sigma_{xx} = \lambda_x \frac{\partial u_x}{\partial x}, \tag{17.3}$$

with λ_x an elastic constant. Assuming constant and homogeneous material properties leads to the one-dimensional wave equation

$$\frac{\partial^2 u_x}{\partial t^2} = \underbrace{\frac{\lambda_x}{\rho}}_{\equiv c_x^2} \frac{\partial^2 u_x}{\partial x^2}, \tag{17.4}$$

with wave speed $c_x = \sqrt{\lambda_x/\rho}$.

17.2 A simple centered second-order scheme in time and space

We saw how to handle the discretization of second-order derivatives in the section devoted to the diffusion equation. For instance, an option for the spatial operator is

$$\frac{\partial^2 u}{\partial x^2} = \frac{1}{\Delta x^2}(u_{k+1} - 2u_k + u_{k-1}) + \mathcal{O}(\Delta x^2). \tag{17.5}$$

Similarly, for the time operator, we can get

$$\frac{\partial^2 u}{\partial t^2} = \frac{1}{\Delta t^2}\left(u^{n+1} - 2u^n + u^{n-1}\right) + \mathcal{O}(\Delta t^2). \tag{17.6}$$

Introduction to Numerical Modeling in Earth and Planetary Sciences, Christian Huber, Oxford University Press.
© Christian Huber (2025). DOI: 10.1093/oso/9780198802716.003.0018

Altogether we can construct an explicit finite difference equation (FDE)

$$u_k^{n+1} = 2u_k^n - u_k^{n-1} + \underbrace{\frac{c_x^2 \Delta t^2}{\Delta x^2}}_{C^2} \left(u_{k+1}^n - 2u_k^n + u_{k-1}^n\right). \tag{17.7}$$

17.2.1 **Implementation**

The FDE (eq. 17.7) yields the stencil shown in Figure 17.1. The second-order differential operator in time introduces a stencil that extends over three different time steps ($n-1$, n and $n+1$). This introduces a challenge when starting the time integration from the initial time, as u_k^{-1} is generally not provided. Most generally the two initial conditions are provided as

$$u(x, t = 0) = f(x) \tag{17.8}$$

$$v(x, t = 0) = \frac{\partial u}{\partial t}|_{t=0} = g(x). \tag{17.9}$$

For the start of the integration, we can use the latter condition to get rid of u_k^{-1} using the approximation

$$\frac{\partial u}{\partial t}|_{t=0} = \frac{u_k^0 - u_k^{-1}}{\Delta t} + \mathcal{O}(\Delta t). \tag{17.10}$$

The time integration is therefore different for the first step

$$u_k^1 = f_k + \Delta t g_k + \underbrace{\frac{c_x^2 \Delta t^2}{\Delta x^2}}_{C^2} \left(u_{k+1}^0 - 2u_k^0 + u_{k-1}^0\right) \tag{17.11}$$

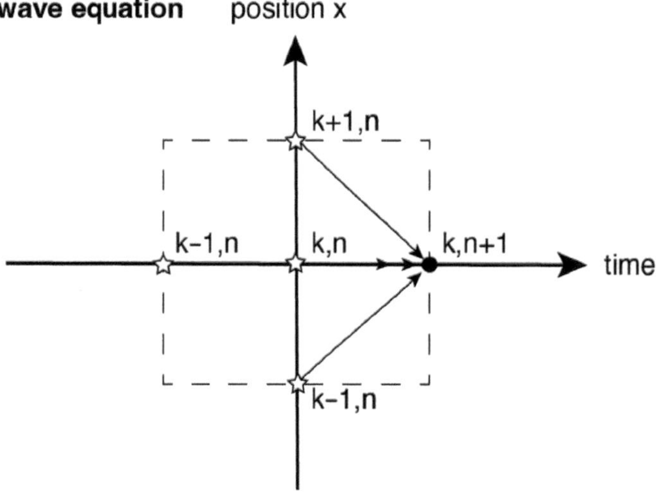

Figure 17.1 Stencil of the second-order wave equation explicit model.

with $f_k = f(k\Delta x)$ and $g_k = g(k\Delta x) = v(k\Delta x, t = 0)$, than for the following steps

$$u_k^{n+1} = 2u_k^n - u_k^{n-1} + \frac{c_x^2 \Delta t^2}{\Delta x^2}\left(u_{k+1}^n - 2u_k^n + u_{k-1}^n\right). \tag{17.12}$$

The stencil of this scheme (beyond the first time step) requires keeping both the previous and penultimate time step in memory to march the equation forward. The pseudo-code appears below.

```
% 1st step
for k=2:Nx-1
        u_new(k)=f(k)+dt g(k)+ ...
                c_x^2 dt^2/ dx^2(u_old(k+1)
                -2u_old(k)+u_old(k-1));
end
u_new(1) =... % BC 1
u_new(Nx) = ... % BC 2
u_older=u_old;
u_old=u_new;
% next time steps
for time=2:maxtime-1 % Marching equation
        for k=2:Nx-1
                u_new(k)=2u_old(k)-u_older(k)+ ...
                        c_x^2 dt^2/ dx^2(u_old(k+1)
                        -2u_old(k)+u_old(k-1));
        end
        u_new(1) =... % BC 1
        u_new(Nx) = ... % BC 2
        u_older=u_old; % u_older is at n-1
        u_old=u_new;
end
```

17.2.2 **Performance**

The first test of the scheme considers a case where the wave motion is triggered by an external forcing that depends on time and space

$$f(x,t) = \left[2c_x^2 - x(L-x)\right]\sin(t). \tag{17.13}$$

The forced one-dimensional wave equation is

$$\frac{\partial^2 u}{\partial t^2} = c_x^2 \frac{\partial^2 u}{\partial x^2} + f(x,t). \tag{17.14}$$

With the addition of the forcing term, the scheme for the first time step becomes

$$u_k^1 = f_k + \Delta t g_k + \underbrace{\frac{c_x^2 \Delta t^2}{\Delta x^2}}_{C^2} \left(u_{k+1}^0 - 2u_k^0 + u_{k-1}^0\right) + \frac{\Delta t^2}{2} f_k^0 \tag{17.15}$$

and the following steps

$$u_k^{n+1} = 2u_k^n - u_k^{n-1} + \frac{c_x^2 \Delta t^2}{\Delta x^2} \left(u_{k+1}^n - 2u_k^n + u_{k-1}^n\right) + \Delta t^2 f_k^n. \tag{17.16}$$

The boundary conditions are set to $u_0^n = u_{Nx+1}^n = 0$. The initial conditions are

$$u(x, t = 0) = \sin\left(\frac{\pi x}{L}\right) \tag{17.17}$$

$$v(x, t = 0) = 0. \tag{17.18}$$

The wave speed $c_x = 0.1$, the length of the domain $L = 1$, and the grid spacing Δx are set to use a Courant number $C = 1$. The wave equation with the applied forcing admits an analytical solution

$$u(x,t) = \sin\left(\frac{\pi x}{L}\right) \cos\left(\frac{\pi c_x t}{L}\right). \tag{17.19}$$

A comparison of the numerical (symbols) and analytical (lines) solutions at different times is provided in the Figure 17.2.

17.2.3 Consistency

The consistency of the scheme is easily demonstrated with (Eq. 17.5) and (Eq.17.6). The truncation error is

$$\mathcal{T} = \underbrace{\frac{\partial^4 u}{\partial x^4} \frac{\Delta x^2}{12} + \mathcal{O}(\Delta x^4)}_{\text{truncation in space } \mathcal{T}_x} + \underbrace{\frac{\partial^4 u}{\partial t^4} \frac{\Delta t^2}{12} + \mathcal{O}(\Delta t^4)}_{\text{truncation in time } \mathcal{T}_t}. \tag{17.20}$$

The scheme is therefore second order in time and space and is not plagued by numerical dispersion.

17.2.4 Stability

Conducting a von Neumann stability analysis with a solution of the type,

$$u_k^n = \sum_j a_j^n \exp(ij\pi \Delta x k) \tag{17.21}$$

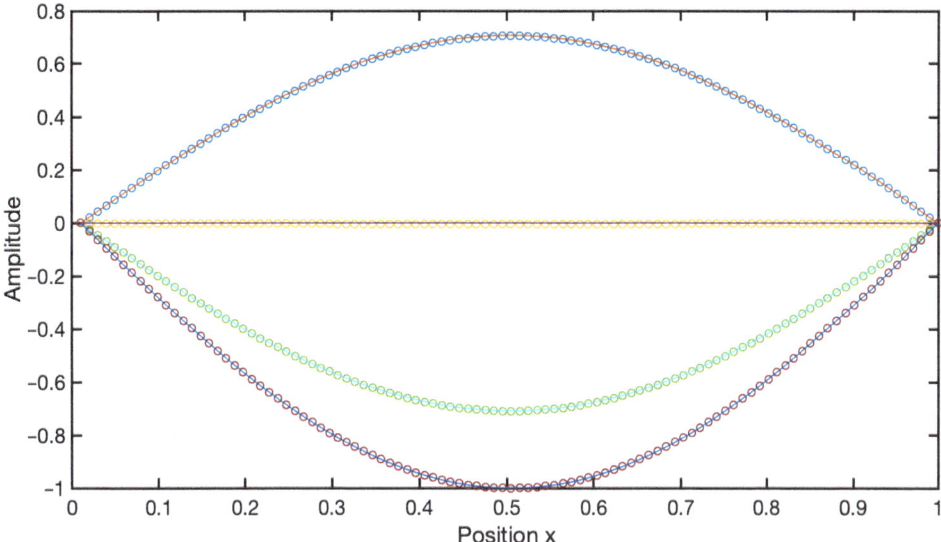

Figure 17.2 Comparison of the numerical and analytical solutions for the wave equation with external forcing discussed in the text. Symbols represent the numerical solution and solid lines the analytical solution at different times.

we obtain from the FDE

$$\sum_j a_j^{n+1} \exp(i\pi j\Delta xk) = 2(1 - C^2)\sum_j a_j^n \exp(i\pi j\Delta xk) + C^2 \sum_j a_j^n \exp(i\pi j\Delta x(k+1))$$
$$+ C^2 \sum_j a_j^n \exp(i\pi j\Delta x(k-1)) - \sum_j a_j^{n-1} \exp(i\pi j\Delta xk). \qquad (17.22)$$

Using the identity

$$2\cos(\pi j\Delta x) = \exp(i\pi j\Delta x) + \exp(-i\pi j\Delta x), \qquad (17.23)$$

and defining

$$\sigma_j = 2\,(1 - \cos(\pi j\Delta x)) \qquad (17.24)$$

we can recast the FDE in the following form

$$\sum_j a_j^{n+1} \exp(i\pi j\Delta xk) = 2\sum_j a_j^n \exp(i\pi j\Delta xk) - C^2 \sum_j \sigma_j a_j^n \exp(i\pi j\Delta x(k+1))$$
$$- \sum_j a_j^{n-1} \exp(i\pi j\Delta xk). \qquad (17.25)$$

Multiplying this equation by $\exp(i\pi j\Delta xm)$ and integrating over a period allows us to retrieve a simple algebraic formulation for the amplitude of the modes in the numerical solution

$$a_m^{n+1} - 2a_m^n - a_m^{n-1} - a_m^n \sigma_m C^2 = (2 - \sigma_m C^2)a_m^n - a_m^{n-1}. \tag{17.26}$$

Simple algebra shows that this equation is equivalent to

$$\begin{pmatrix} a_m^n \\ a_m^{n+1} \end{pmatrix} = \overbrace{\begin{pmatrix} 0 & 1 \\ -1 & \underbrace{2 - \sigma_m C^2}_{\equiv a} \end{pmatrix}}^{\mathcal{M}} \begin{pmatrix} a_m^{n-1} \\ a_m^n \end{pmatrix}. \tag{17.27}$$

We see that the vector of amplitude is rescaled at each iteration by a matrix $\mathcal{M}$. How does the matrix $\mathcal{M}$ rescale the amplitude? This can be quantified by computing the eigenvalues λ of $\mathcal{M}$. If the eigenvalues are such that $|\lambda| \leq 1$ then the amplitudes are not growing over time and the scheme is stable. The eigenvalues are

$$\lambda_{\pm} = \frac{-a \pm \sqrt{a^2 - 4}}{2}. \tag{17.28}$$

Then

$$|\lambda| \leq 1 \quad \rightarrow \quad -2 \leq \underbrace{\sigma_m C^2 - 2}_{=a} \leq 2. \tag{17.29}$$

Knowing that

$$0 \leq \sigma_m \leq 4. \tag{17.30}$$

These two conditions reduce to the Courant–Friedrichs–Lewy (CFL) condition

$$C \leq 1. \tag{17.31}$$

The explicit finite difference model for the wave equation is then conditionally stable.

17.3 **Wave propagation in one-dimensional heterogeneous media**

One common application of wave propagation is to constrain heterogeneities in elastic properties or density in the subsurface. This section introduces a modified version of the wave equation scheme presented in Section 17.2 to account for heterogeneities in elastic properties. Starting back from (Eq. 17.4), but now assuming that $\lambda_x(x)$ (for simplicity we still assume density to be constant and homogeneous), the wave equation takes the form

$$\frac{\partial^2 u}{\partial t^2} = \frac{\partial}{\partial x}\left(c(x)^2 \frac{\partial u}{\partial x}\right), \tag{17.32}$$

with wave speed $c_x = \sqrt{\lambda_x(x)/\rho}$. Let us now define a function q related to the elastic stress (stress/density)

$$q(x) = c(x)^2 \frac{\partial u}{\partial x}, \tag{17.33}$$

the wave equation reduces then to

$$\frac{\partial^2 u}{\partial t^2} = \frac{\partial q}{\partial x}. \tag{17.34}$$

17.3.1 **Implementation**

As q relates directly to the stress and that stress needs to be continuous and balanced across any interface, we introduce the notation

$$q_{k\pm1/2} = q((k \pm 1/2)\Delta x), \tag{17.35}$$

which relates to the values of the stress at the mid-point between consecutive nodes, that is, at the interface between cells centered on the consecutive nodes themselves (see Figure 17.3 for a schematic illustration).

We define an approximation of the spatial derivative of q at position x_k

$$\frac{\partial q}{\partial x} = \frac{q_{k+1/2} - q_{k-1/2}}{\Delta x} + \mathcal{O}(\Delta x). \tag{17.36}$$

Furthermore, we define

$$q_{k+1/2} = c^2_{k+1/2}\frac{u_{k+1} - u_k}{\Delta x} \tag{17.37}$$

$$q_{k-1/2} = c^2_{k-1/2}\frac{u_k - u_{k-1}}{\Delta x}. \tag{17.38}$$

The approach we take here is often referred to as a flux-conservative approach (see [17]) and it is preferential to using the chain rule for the product $c_x \partial u/\partial x$ because it preserves momentum balances locally even when material properties are not homogeneous. The last step before implementing this scheme is to define the interpolated values of $c(x)$ at the midpoint between nodes. Different averaging schemes are possible, the harmonic mean is well-suited to deal with abrupt changes in properties, and we will use it here

$$c_{k\pm1/2} = 2\left[\frac{1}{c_k} + \frac{1}{c_{k\pm1}}\right]^{-1}. \tag{17.39}$$

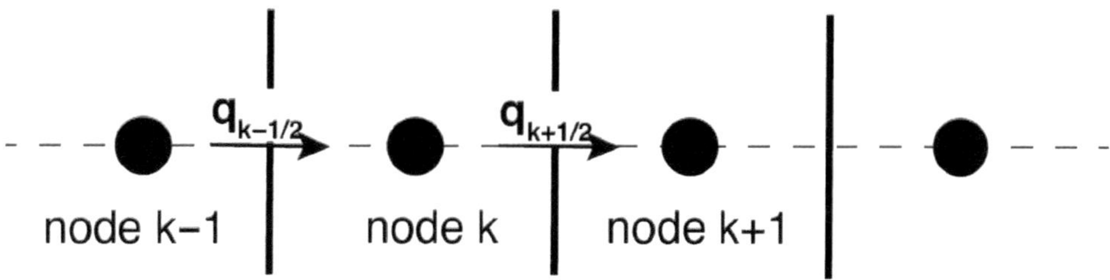

Figure 17.3 Illustration of the flux conservative scheme for the wave equation. Fluxes are computed at interfaces located at the midpoint between adjacent gridpoints.

The modified pseudo-code for this model is

```
% 1st step
for k=2:Nx-1
        c_p(k)=2(1/c(k)+1/c(k+1))^(-1);
        c_m(k)=2(1/c(k)+1/c(k-1))^(-1);
        q_p(k)=c_p(k)^2*(u_old(k+1)-u_old(k));
        q_m(k)=c_m(k)^2*(u_old(k)-u_old(k-1));

        u_new(k)=u_old(k)+dt g(k)+ ...
                c_x^2 dt^2/ dx(q_p(k)-q_m(k));
end
c_p(1)=...
c_m(1)=...
q_p(1)=...
q_m(1)=... % BC 1
u_new(1)=...
 %% repeat same last 5 lines for k=Nx
...

u_older=u_old;
u_old=u_new;
% next time steps
for time=2:maxtime-1 % Marching equation
        for k=2:Nx-1
                q_p(k)=c_p(k)^2*(u_old(k+1)-u_old(k));
                q_m(k)=c_m(k)^2*(u_old(k)-u_old(k-1));
                u_new(k)=2u_old(k)-u_older(k)+ ...
                        c_x^2 dt^2/ dx(q_p(k)-q_m(k));
        end
        q_p(1)=...
        q_m(1)=...
        u_new(1)=...
        q_p(Nx)=...
        q_m(Nx)=...
        u_new(Nx)=...
        u_older=u_old; % u_older is at n-1
        u_old=u_new;
end
```

17.3.2 Performance

We consider a two-layer medium with $0 \leq x \leq L$ and

$$c(x) = \begin{cases} 0.3 & \text{if } x < 0.6L, \\ 0.1 & \text{otherwise.} \end{cases} \tag{17.40}$$

The initial conditions are

$$u(x, t = 0) = \exp\left(-\frac{x^2}{\sigma^2}\right) \tag{17.41}$$

$$v(x, t = 0) = \frac{\partial u}{\partial t}|_{t=0} = 0. \tag{17.42}$$

The Courant number $C = \max(c(x))\Delta t/\Delta x$ is set to 1. The left and right boundary conditions are $q_{1/2} = 0$ and $q_{Nx+1/2} = 0$, respectively. Snapshots of the numerical solution are shown in Figure 17.4. Interestingly, half of the initial perturbation energy is moving to the left (outside of the domain), which explains why the amplitude of the incident wave (moving to the right) is only 0.5. As the incident wave reaches the boundary between the different material properties (shaded in gray) the incident wave produces a reflected and a transmitted wave. From theory, the amplitude of the reflected and transmitted waves are related to that of the incident wave through reflection (R) and transmission (T) coefficients

$$R = \frac{Z_1 - Z_2}{Z_1 + Z_2} \tag{17.43}$$

$$T = 1 + R, \tag{17.44}$$

where $Z_{1,2}$ refers to the impedance ($Z = \rho c_x$) of the two layers. As we assume the same density everywhere, the impedance contrast reduces to the wave speed contrast between the two layers and for the choice of wave speeds we expect $R = 0.5$ and $T = 1.5$, which lead to amplitudes of 0.25 and 0.75 for the reflected and transmitted waves, respectively. This is in very good agreement with the numerical results.

17.4 A simple centered second-order scheme in time (implicit) and space

An implicit version of the previous scheme yields

$$u_k^{n+1} = 2u_k^n - u_k^{n-1} + \underbrace{\frac{c_x^2 \Delta t^2}{\Delta x^2}}_{C^2} \left(u_{k+1}^{n+1} - 2u_k^{n+1} + u_{k-1}^{n+1}\right). \tag{17.45}$$

The stencil for this scheme is different and involves three unknowns per FDE (value of k). A matrix vector system of equation is therefore required to solve simultaneously for the displacement at all

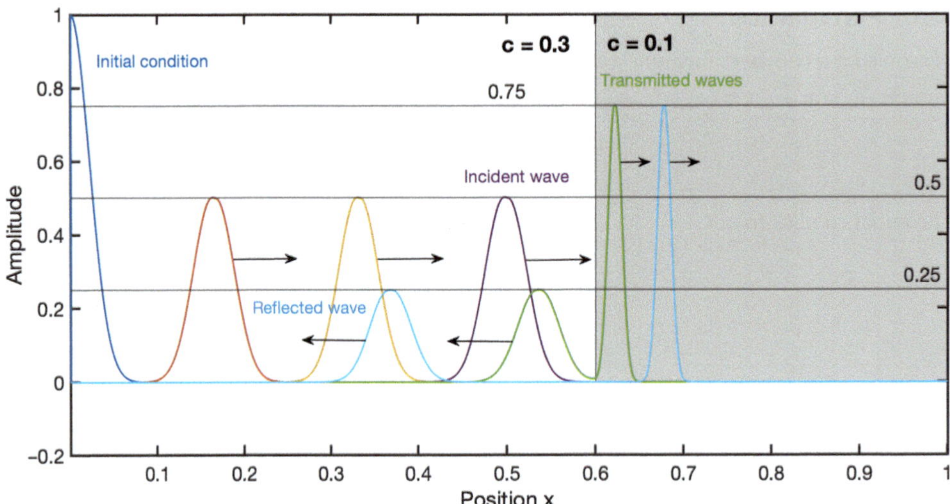

Figure 17.4 One-dimensional wave propagation in a layered medium. The different lines show snapshots taken at different times. The initial perturbation (half Gaussian) is centered at $x = 0$. The incident wave travels to the right with an amplitude of 0.5 before interacting with the interface at $x = 0.6L$, where it gives way to a reflected wave (amplitude of 0.25) and a transmitted wave (0.75).

positions. Under these circumstances, defining a displacement array

$$\mathbf{u}^n = \begin{pmatrix} u_1^n \\ u_2^n \\ \vdots \\ u_{Nx}^n \end{pmatrix} \tag{17.46}$$

we can build a marching equation of the form

$$\mathbf{M}\mathbf{u}^{n+1} = \mathbf{R}. \tag{17.47}$$

The matrix $\mathbf{M}$ is of dimensions $Nx \times Nx$

$$\mathbf{M} = \begin{pmatrix} 1+2C^2 & -C^2 & 0 & \dots & & & \\ -C^2 & 1+2C^2 & -C^2 & 0 & \dots & & \\ 0 & -C^2 & 1+2C^2 & -C^2 & 0 & \dots & \\ \vdots & & & & & & \\ & & \dots & 0 & -C^2 & 1+2C^2 & \\ & & & & & & \end{pmatrix}, \tag{17.48}$$

which again is tridiagonal and can be inverted by a Thomas algorithm, for example. So far the solution procedure resembles that of the implicit diffusion scheme, although the right-hand side of the marching equation is different as it involves more terms and includes displacement at previous iterations. A reorganization of (Eq. 17.45) shows that the element k of the array $\mathbf{R}$ becomes

$$R_k = 2u_k^n - u_k^{n-1}. \tag{17.49}$$

Boundary conditions are again required to modify the array **R** for the first and last rows. Starting with Dirichlet-type conditions, if the displacement at u_0 is given by u_{left} and $u_{Nx+1} = u_{right}$, then

$$R_1 = 2u_1^n - u_1^{n-1} + C^2 u_{left} \tag{17.50}$$

$$R_{Nx} = 2u_{Nx}^n - u_{Nx}^{n-1} + C^2 u_{right}. \tag{17.51}$$

17.4.1 **Implementation**

Similarly to the explicit scheme described earlier, the first time step needs to be handled with care so that the two sets of initial conditions (on $u(t=0,x)$ and $v(t=0,x) = \frac{\partial u}{\partial t}(t=0,x)$ are used. The pseudo-code appears below.

```
u_ini=... % initial conditions
v=.. % velocity ini conditions
build matrix M...
invM=inv(M);
R=u_ini-v;
R(1)=R(1)+C^2*u_left;
R(Nx)=R(Nx)+C^2*u_right;
u_old=u_ini;
% 1st step
for n=1:Niter
        u_new=invM*R;
        u_older=u_old;
        u_old=u_new;
        R=2*u_old-u_older;
        R(1)=R(1)+C^2*u_left;
        R(Nx)=R(Nx)+C^2*u_right;
end
```

Obviously a different set of boundary conditions would require modifications in the matrix **M** and array **R**, but the approach is similar for flux (here strain) boundary conditions to the flux boundary conditions presented earlier for the diffusion implicit model.

17.4.2 **Consistency**

While the FDE scheme and stencil are different from the explicit model, the truncation error is largely similar; it is second order in time and space because of the centered difference schemes.

The scheme is naturally consistent. Akin to the diffusion equation, the main difference between explicit and implicit lies in the stability of the scheme.

17.4.3 **Stability**

Repeating the von Neumann stability analysis with a solution of the type

$$u_k^n = \sum_j a_j^n \exp(ij\pi\Delta xk), \tag{17.53}$$

we obtain a slightly different equation than for the explicit scheme

$$\begin{aligned}\sum_j a_j^{n+1}\exp(i\pi j\Delta xk) &= 2\sum_j a_j^n \exp(i\pi j\Delta xk) - 2C^2\sum_j a_j^{n+1}\exp(i\pi j\Delta xk)\\ &+ C^2\sum_j a_j^{n+1}\exp(i\pi j\Delta x(k+1))\\ &+ C^2\sum_j a_j^{n+1}\exp(i\pi j\Delta x(k-1)) - \sum_j a_j^{n-1}\exp(i\pi j\Delta xk).\end{aligned} \tag{17.54}$$

Using the identity

$$2\cos(\pi j\Delta x) = \exp(i\pi j\Delta x) + \exp(-i\pi j\Delta x), \tag{17.55}$$

and defining

$$\sigma_j = 2\,(1 - \cos(\pi j\Delta x)) \tag{17.56}$$

with $0 \leq \sigma_j \leq 4$, we can recast the FDE into the following form

$$\begin{aligned}\sum_j a_j^{n+1}\exp(i\pi j\Delta xk) &= 2\sum_j a_j^n \exp(i\pi j\Delta xk) - C^2\sum_j \sigma_j a_j^{n+1}\exp(i\pi j\Delta x(k+1))\\ &- \sum_j a_j^{n-1}\exp(i\pi j\Delta xk).\end{aligned} \tag{17.57}$$

Multiplying this equation by $\exp(i\pi j\Delta xm)$ and integrating over a period allows us to retrieve a simple algebraic formulation for the amplitude of the modes in the numerical solution

$$\left(1 + \sigma_m C^2\right) a_m^{n+1} = 2a_m^n - a_m^{n-1}. \tag{17.58}$$

Simple algebra shows that this equation is equivalent to

$$\begin{pmatrix} a_m^n \\ a_m^{n+1} \end{pmatrix} = \overbrace{\begin{pmatrix} 0 & 1 \\ -\frac{1}{1+\sigma_m C^2} & \frac{2}{1+\sigma_m C^2} \end{pmatrix}}^{\mathcal{M}} \begin{pmatrix} a_m^{n-1} \\ a_m^n \end{pmatrix}. \tag{17.59}$$

The eigenvalues λ of $\mathcal{M}$, control the stability of the solution, and the scheme is stable if the eigenvalues are such that $|\lambda| \leq 1$. The eigenvalues are

$$\lambda_\pm = \frac{2 \pm \sqrt{4 - 4a}}{2a}, \tag{17.60}$$

where $a = 1 + \sigma_m C^2 \geq 1$. Then multiplying the eigenvalue with its complex conjugate yields

$$|\lambda_\pm| = \frac{1}{\sqrt{a}} \leq 1. \tag{17.61}$$

This shows that the implicit scheme is unconditionally stable.

17.5 **Exercises**

- Implement the case presented in Figure 17.4. Vary the coefficients of reflection and transmission to test different cases and compare the amplitude of the reflected and transmitted waves with expected values for different cases.
- Implement a homogeneous wave propagation model (homogeneous density and elastic moduli) with an initial excitation and boundary conditions with no displacement at both ends of the domain. Be careful! The initial perturbation needs to be compatible with these boundary conditions! Define the initial perturbation to find standing waves of different wavelengths (from twice the domain for the longest to two wavelengths for the smaller). Measure if you have any dissipation of energy in the system (from the amplitude change over different cycles). What could be the cause of these changes? How can you minimize them given the starting equation does not predict any damping?

Chapter 18

The shallow water equation

The shallow water equation is a nonlinear partial differential equation (PDE) that describes the motion of water within a basin under the assumption that waves propagating at the surface have a wavelength significantly longer than the depth of the basin. The equations give rise to gravity waves, which propagate with a velocity proportional to $(gh)^{1/2}$ where h is the amplitude of the wave. These equations are often used to model tsunamis, tides, and sometimes Rossby or Kelvin waves in the ocean or atmosphere. One of the key assumption is that the fluid is inviscid, that is, dissipation caused by the fluid's viscosity is neglected. Under these assumptions, the waves are nondispersive.

18.1 Derivation of the nonlinear partial differential equation (PDE)

The shallow water equation describes mass and momentum conservation considering a vertical section of water in a basin. Its derivation is founded on several assumptions that allow for simplifications of the equations of motion leading to an hyperbolic system of PDEs. Here we focus our attention on a one-dimensional expression (waves traveling along the direction x). The set of starting conservation equations is

$$\nabla \cdot \mathbf{v} = 0 \tag{18.1}$$

$$\frac{\partial \mathbf{v}}{\partial t} + \mathbf{v} \cdot \nabla \mathbf{v} + \frac{1}{\rho} \nabla p + g\hat{\mathbf{z}} = 0, \tag{18.2}$$

where ρ, p, g and $\mathbf{v}$ are respectively the water density, pressure, acceleration due to gravity, and the velocity field. The first equation states that water is assumed incompressible and the second is a statement of momentum balance and neglects viscous stresses (inviscid approximation). A third equation is required to describe the free surface boundary condition at the water–atmosphere interface (here surface tension is neglected) and reads

$$\frac{\partial \eta}{\partial t} + \mathbf{v} \cdot \nabla \eta = v_z, \quad \text{at } z = \eta(x, t) \tag{18.3}$$

with η being the vertical position of the water interface with reference to the water level at rest and v_z the z-component of the velocity field. For readers used to fluid dynamical concepts the latter equation states that the Lagrangian (material) derivative of the position of the free interface matches the local vertical component of the velocity field.

Defining the depth of the water column at rest to be h and integrating the mass conservation equation over the depth of the full water column $-h \leq z \leq \eta$, we obtain the one-dimensional

Introduction to Numerical Modeling in Earth and Planetary Sciences, Christian Huber, Oxford University Press.
© Christian Huber (2025). DOI: 10.1093/oso/9780198802716.003.0019

approximation for the mass conservation

$$\frac{\partial \eta}{\partial t} + \frac{\partial}{\partial x}\left[(\eta + h)\, u\right] = 0, \tag{18.4}$$

where u is the depth-averaged horizontal component of the velocity field and we made use of the boundary condition $\mathbf{v} = 0$ at the water–substrate interface ($z = -h$) as well as the other boundary condition (Eq. 18.3).

Simplifying the momentum equation with the assumption that the flow is dominantly in the x direction and assuming that the horizontal wavelength of the surface perturbation η is significantly greater than the depth of the water column h, we get

$$\frac{\partial u}{\partial t} + u\frac{\partial u}{\partial x} + g\frac{\partial \eta}{\partial x} = 0. \tag{18.5}$$

If we further assume here that the water–substrate interface is horizontal (constant depth h) and define $H = \eta + h$, then the mass and momentum conservation become

$$\frac{\partial H}{\partial t} + u\frac{\partial H}{\partial x} + H\frac{\partial u}{\partial x} = 0 \tag{18.6}$$

$$\frac{\partial u}{\partial t} + u\frac{\partial u}{\partial x} + g\frac{\partial H}{\partial x} = 0. \tag{18.7}$$

Note that these are hyperbolic nonlinear PDEs because of products $u\frac{\partial H}{\partial x}$, $H\frac{\partial u}{\partial x}$, and $u\frac{\partial u}{\partial x}$.

18.2 A simple scheme for the shallow water equation

The scheme we look at is pseudo-implicit (time) and uses centered difference approximations for the first derivative over space. The finite difference equation (FDE) for the mass conservation above is

$$\frac{H_k^{n+1} - H_k^n}{\Delta t} + \frac{u_k^n}{2\Delta x}\left[H_{k+1}^{n+1} - H_{k-1}^{n+1}\right] + \frac{H_k^n}{2\Delta x}\left[u_{k+1}^{n+1} - u_{k-1}^{n+1}\right] = 0. \tag{18.8}$$

Similarly we define the FDE for the momentum conservation

$$\frac{u_k^{n+1} - u_k^n}{\Delta t} + \frac{u_k^n}{2\Delta x}\left[u_{k+1}^{n+1} - u_{k-1}^{n+1}\right] + \frac{g}{2\Delta x}\left[H_{k+1}^{n+1} - H_{k-1}^{n+1}\right] = 0. \tag{18.9}$$

We note that the FDEs are basically implicit, except for the pre-factors multiplying the spatial derivatives in the nonlinear term, which are considered at the time step n instead of $n + 1$ (pseudo-implicit assumption). This linearization assumption in our discretization allows us to gather the algebraic equations above in a matrix–vector form.

18.3 **Test simulation setup**

We test the implementation of the shallow water equations with a simple problem where the nonperturbed depth of the water layer is set to $h = 0.05$. The surface perturbation is initially set to

$$\eta = \frac{2h}{5}\exp(-5x^2), \tag{18.10}$$

which yields an initial condition

$$H(x, t = 0) = h\left(1 + \frac{2}{5}\exp(-5x^2)\right). \tag{18.11}$$

In the simulation the gravity constant $g = 1$ and we use 200 nodes to cover the domain ($Nx = 200$). The Courant number C is set so that it is unity for the velocity $U_{max} = \sqrt{\frac{7}{5}hg}$, which corresponds to the maximum velocity given the initial perturbation for a critical flow ($Fr = U/\sqrt{gh} = 1$).

Figure 18.1a, b shows snapshots in time of the water level and horizontal velocity at different time as the perturbation progresses and produces a wave at the surface. As time progresses the front of the wave becomes steeper, but now we clearly see dispersion. An analysis of the relation between water depth and flow velocity defines a clear trend that is compared to the trends expected for different fixed values of the Froude number. The flow in this calculation is subcritical ($Fr < 1$).

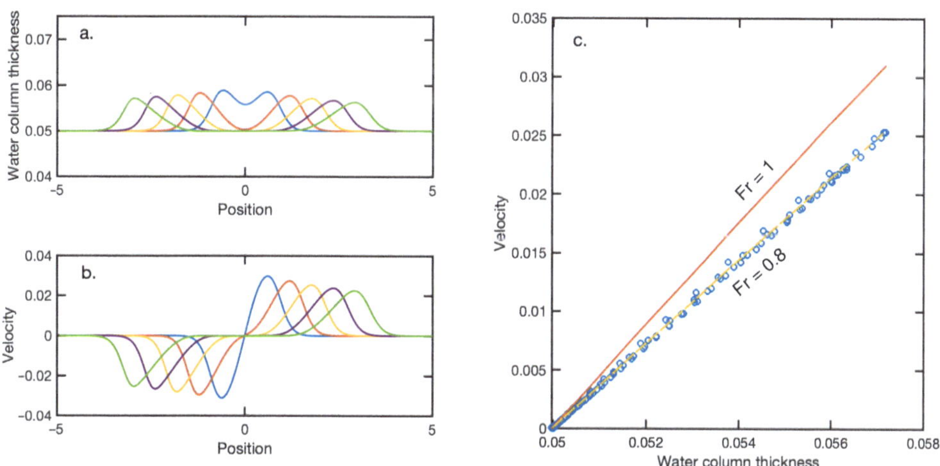

Figure 18.1 Snapshots of the numerical solution to the shallow water equation. After the initial perturbation, waves propagate away on both sides with fronts steepening over time. Relating the horizontal speed u to the depth of the water column gives rise to a simple correlation that depends on the Froude number of the flow. Here the flow is subcritical ($Fr < 1$), which explains why we do not see a step-like instability forming at the front of the perturbation.

18.4 Exercises

- Reproduce the example shown in Figure 18.1.
- How would you change the conditions to get a critical flow ($Fr = 1$) or even a supercritical flow ($Fr > 1$)?

Part IV

Overview of other numerical methods

Chapter 19

Top-down approaches

The finite difference methods discussed at length in this book are but one of the top-down approaches that are commonly used to solve differential equations numerically. The top-down qualification refers to methods that start from a continuum mathematical description (differential equations with continuous fields) and generally seek to reduce the differential equations to a set of algebraic equations through spatial and temporal discretization. While the finite difference method is perhaps the most intuitive, as it takes the differential operators and finds a direct discretization through Taylor series truncation, other methods exist and present their own advantages and limitations. While not the focus of this book, they deserve a rapid introduction for the interested reader, and references are mentioned to guide readers toward further exploration.

19.1 Finite volume methods

The finite volume approach shares many similarities with finite differences and, in certain circumstances, yields similar and even sometimes identical algebraic equations. The main difference between the two approaches is that finite volumes starts from the integral version of the conservation law, rather than the generic differential form (as does finite differences approach). For illustration, we consider a simple statement of a steady one-dimensional heat conduction with heat sources $f(x)$ and heat flux $q(x) = -k(x)\frac{dT}{dx}$. Originally the ordinary differential equation (ODE) for this problem is

$$\frac{dq}{dx} = f(x). \tag{19.1}$$

We now define control volumes centered here on the nodes at position x_i. Each control volume is bounded by faces at the half point between nodes (see Figure 19.1). The idea of the finite volume method is to integrate the conservation law over the domain, and more specifically over each control volume (domain represented by Ω) leading to

$$\int_\Omega \frac{dq}{dx}\,\mathrm{d}\Omega = \int_\Omega f(x)\mathrm{d}\Omega. \tag{19.2}$$

Note that considering more than one spatial dimension would just lead to an analogous statement

$$\int_\Omega \nabla\cdot\vec{q}\,\mathrm{d}\Omega = \int_\Omega f(x)\mathrm{d}\Omega. \tag{19.3}$$

Introduction to Numerical Modeling in Earth and Planetary Sciences, Christian Huber, Oxford University Press.
© Christian Huber (2025). DOI: 10.1093/oso/9780198802716.003.0020

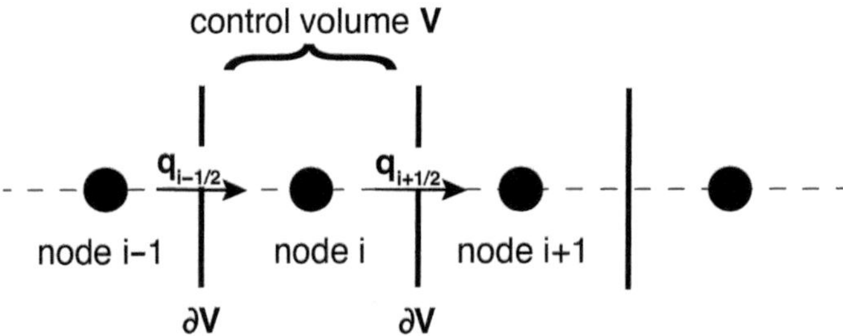

Figure 19.1 Schematic illustration of a finite volume one-dimensional regular grid. Note that the nodes here are located at the center of the control volumes, which are bounded by interfaces ∂V. Fluxes (here $q_{i\pm1/2}$) are computed at midpoints between nodes at the control volume interfaces.

We can use the divergence theorem or the fundamental theorem of calculus for the one-dimensional problem to reduce the integral of the derivative of the heat flux into fluxes at the faces of the control volume

$$\int_{\partial\Omega} \vec{q} \cdot \hat{n} \, \mathrm{d}\partial\Omega = \int_{\Omega} f(x) \mathrm{d}\Omega, \tag{19.4}$$

where $\partial\Omega$ refers to the boundary of the control volume and $\hat{n}$ is the normal (here defined positive into the control volume) boundary (see Figure 19.1).

Assume f_i to be the volume-averaged value of $f(x)$ within the control volume associated with node i, $q_{i+1/2}$ and $q_{i-1/2}$ to be the interface (surface) averaged values for the fluxes crossing respectively the interface between the control volumes $i+1$ and i, and i and $i-1$. We can integrate equation 19.4 to get

$$(q_{i-1/2} - q_{i+1/2}) \Delta A = f_i \Delta x \Delta A, \tag{19.5}$$

where ΔA and Δx are respectively the surface area of the interfaces and the distance between successive interfaces ($\Delta x \Delta A$ is effectively the volume of the control volume). The definition of the conductive heat flux

$$q = -k \frac{dT}{dx}, \tag{19.6}$$

allows for different choices of approximations for $q_{i\pm1/2}$. For simplicity we use a linear approximation

$$q_{i-1/2} = -k_{i-1/2} \left(\frac{T_i - T_{i-1}}{\Delta x} \right) \tag{19.7}$$

$$q_{i+1/2} = -k_{i+1/2} \left(\frac{T_{i+1} - T_i}{\Delta x} \right) \tag{19.8}$$

and the thermal conductivities using a linear interpolation are

$$k_{i-1/2} = \frac{k_i + k_{i-1}}{2} \tag{19.9}$$

$$k_{i+1/2} = \frac{k_i + k_{i+1}}{2}. \tag{19.10}$$

After reorganization we obtain an algebraic equation for the finite volume i

$$\frac{1}{\Delta x}\left(k_{i+1/2}\frac{T_{i+1} - T_i}{\Delta x} - k_{i-1/2}\frac{T_i - T_{i-1}}{\Delta x}\right) = f_i. \tag{19.11}$$

This should be familiar now as it is similar to what we would get from a centered finite difference scheme. Now collecting algebraic equations for each control volume together allows us to recast the finite volume scheme in terms of the matrix–vector equation

$$\mathcal{M}\vec{T} = \vec{f}, \tag{19.12}$$

which can be inverted to retrieve the distribution of temperature along the domain.

This simple example is provided to illustrate differences and similarities between finite difference and finite volume methods. In this particular case (i.e., structured regular grid, one dimension) the two methods yield very similar algebraic equations. Again the main conceptual difference is that the finite volume approach deals with the integral version of the governing equation. By design, finite volume methods perfectly conserve mass, momentum, or energy both locally (at the scale of the control volume) and over the whole domain, while this is not guaranteed for finite difference methods. This does not mean that the results are always correct with finite volume solutions, that is, convergence is not guaranteed, as stability limitations exist in a very similar way as for finite difference methods. For more information about the finite volume method and some applications, interested readers are encouraged to look at [17].

19.2 **Finite element methods**

The finite element method (FEM) is a mesh-based approach, akin to the finite difference and finite volume approaches and it is generally considered a specific application of Galerkin methods. The strategy of FEM is to take the differential equation, project it onto a set of test functions, and decompose them onto a set of basis functions to retrieve, yet again, a set of algebraic equations. An important aspect of FEM is that it uses also an integral (weak) form of the differential equations, which presents some advantages. As described in Part I, the existence of derivatives requires continuity of the fields, so it is possible that the fields may, in reality, be discontinuous at some locations (e.g., localized phase change). In that case, presenting the governing equations in integral form is preferential as we can integrate functions that contain discontinuities (under some conditions). There are some similarities with finite volume methods as the differential equations are integrated over *elements* (control volumes). Again we start from a steady conduction problem

$$\nabla \cdot \vec{q} = f(\vec{x}), \tag{19.13}$$

where $\vec{q}$ is the conductive heat flux and $f(\vec{x})$ is a source term. The first step is to define a residual $r(\vec{x})$ such that

$$\nabla \cdot \vec{q} - f(\vec{x}) = r(\vec{x}). \tag{19.14}$$

Introducing now a test function $v(\vec{x})$ (more on that shortly) and multiplying both sides with it and integrating over a domain Ω, which can represent the whole domain or an individual element, we get a weak form description of the problem

$$\int_{\Omega} (\nabla \cdot q - f(\vec{x}))\, v(\vec{x})\, \mathrm{d}\Omega = \int_{\Omega} r(\vec{x}) v(\vec{x})\, \mathrm{d}\Omega. \tag{19.15}$$

If (Eq. 19.15) is null for any choice of test function $v(x)$, then it means that the residual $r(\vec{x}) = 0$ and we have found the solution to the problem. Using

$$\nabla \cdot (\vec{q} v) = v \nabla \cdot \vec{q} + \nabla v \cdot \vec{q}, \tag{19.16}$$

the governing equation to satisfy is now

$$\int_{\Omega} [\nabla \cdot (\vec{q} v) - \nabla v \cdot \vec{q}]\, \mathrm{d}\Omega - \int_{\Omega} f v\, \mathrm{d}\Omega = 0. \tag{19.17}$$

Using the divergence theorem on the first term leads to

$$\int_{\partial\Omega} v\vec{q} \cdot \hat{n}\, \mathrm{dS} - \int_{\Omega} \nabla v \cdot \vec{q}\, \mathrm{d}\Omega - \int_{\Omega} f v\, \mathrm{d}\Omega = 0. \tag{19.18}$$

Using now that the conductive flux $\vec{q} = -k \nabla T$ or in one dimension $q = -k\frac{dT}{dx}$ and sticking with one dimension for simplicity, and after a little bit of reorganization

$$\int_{\Omega} \frac{dv}{dx} k \frac{dT}{dx}\, \mathrm{dx} = \int_{\Omega} f v\, \mathrm{dx} - vq|_{\partial\Omega}, \tag{19.19}$$

where the left side depends on both T, v and the right side only on v. We now introduce the basis functions $\phi_k(\vec{x})$ (different choices can be used depending on the complexity and accuracy targeted) such that

$$T(\vec{x}) = \sum_j a_j \phi_j(\vec{x}) \tag{19.20}$$

$$v(\vec{x}) = \sum_i b_i \phi_i(\vec{x}). \tag{19.21}$$

Introducing these definitions into the weak form

$$\int_{\Omega} \sum_i b_i \frac{d\phi_i}{dx} k \sum_j a_j \frac{d\phi_j}{dx}\,\mathrm{dx} = \int \Omega f \sum_i b_i \phi_i(\vec{x})\,\mathrm{dx} - \sum_i b_i \phi_j q(\partial\Omega). \tag{19.22}$$

After further reorganization (coefficients a, b do not depend on position) it is easy to show that this equation reduces to

$$\sum_{i,j} b_i \left(K_{ij} a_j - R_j\right) = 0, \tag{19.23}$$

where

$$K_{ij} = \int_{\Omega} \frac{d\phi_i}{dx} k \frac{d\phi_j}{dx}\,\mathrm{dx} \tag{19.24}$$

$$R_j = \int_{\Omega} \phi_j(\vec{x}) f(\vec{x})\,\mathrm{dx} + \phi_j(\vec{x}) q(\partial\vec{\Omega}). \tag{19.25}$$

Remembering that this should be true for any test function $v(\vec{x})$, and therefore any choice of sets of b_i values, the FEM problem reduces to the algebraic system of equation

$$K\vec{a} = \vec{R}, \tag{19.26}$$

which can be solved by inverting the matrix K to get the coefficients a_j of $T(\vec{x})$. This may not be trivial as the matrix can be large and not sparse. Often this inversion is not computed in a brute force way, but iteratively as a suite of approximation (e.g., using a conjugate gradient algorithm). Note that the choice of basis function ϕ informs the derivatives and definitions of K and R together with the source term f and the boundary conditions $q(\partial\Omega)$. Similar to finite volume methods, FEM provides a method where conserved quantities are perfectly conserved locally (i.e., each element) and globally (i.e., the whole domain). The weak form also provides the advantage that the method can deal with discontinuities (e.g., material properties such as the conductivity). Obviously, the basis functions need to be differentiable over each element, which precludes these discontinuities from being located within an element, but clever meshing allows us to match discontinuities with interfaces between elements to overcome this issue. Interested readers are directed to a large literature on FEM. One particular reference that is quite effective at giving a concise but complete description of the method is Zohdi [28].

Chapter 20

Bottom-up approaches

In contrast with top-down approaches, which directly transform the differential equations into a set of algebraic equations through different approximations, bottom-up approaches generally consider the physical problem at a smaller scale (below continuum scale) where first principle physics prevails (Newton's law). Statistical approaches are required to upscale model results to the continuum scale. There are several distinct methods that qualify here as bottom-up approaches, such as molecular dynamics and other particle based methods. We discuss one of them: the lattice Boltzmann method.

20.1 Lattice Boltzmann method

The lattice Boltzmann method is rooted in kinetic theory (a branch of statistical mechanics). The main idea is that if we consider a continuum as a collection of "particles" (here generally not to be confused with atoms or molecules like in molecular dynamics) that interact through simple rules, then it should be possible to tune the particle model to retrieve the desired continuum equations through statistical averaging. While the benefit of such an approach is that the physics reduces to Newton's second law and is therefore simple to handle, the number of particles that need to be tracked to yield satisfying statistics to derive continuum-scale model predictions exceeds what can be handled even with modern computers. This is where statistical mechanics comes in. Specifically, more than a century ago, Ludwig Boltzmann had a the brilliant idea (yet controversial at the time) of defining probability distribution functions $f(\vec{x}, \vec{v}, t)$ describing the probability of a particle to be located at time t at position $\vec{x}$ and traveling with the instantaneous velocity $\vec{v}$. Note that f is not only a function of time and space (position), but also that it depends on the (vectorial) velocity of the particle. A simple discrete conservation statement for this probability function is

$$f(\vec{x} + \vec{v}\Delta t, \vec{v}, t + \Delta t) - f(\vec{x}, \vec{v}, t) = \Omega(\vec{x}, \vec{v}, t). \tag{20.1}$$

(Equation 20.1) implies that, in the absence of collisions that would redistribute momentum, $\Omega(\vec{x}, \vec{v}, t) = 0$ because the probability of a free streaming particle should remain unchanged. In other words, if no net force is applied to a particle traveling with velocity $\vec{v}$ at time t and located at $\vec{x}$, then at $t + \Delta t$ it should be located at $\vec{x} + \vec{v}\Delta t$ with the same velocity. This probability function is a continuous field and therefore this could be written as (in the reference frame of the moving particles)

$$\frac{Df(\vec{x}, \vec{v}, t)}{Dt} = 0, \tag{20.2}$$

where the material derivative is implied. Again this is not generally true because a force field and particle–particle interactions (such as collisions) would provide a source term to that equation (Ω is thereafter referred to a collision operator and external forces can be added similarly).

Introduction to Numerical Modeling in Earth and Planetary Sciences, Christian Huber, Oxford University Press.
© Christian Huber (2025). DOI: 10.1093/oso/9780198802716.003.0021

The main idea of the lattice Boltzmann method is to take the discrete version of Boltzmann's equation (Eq. 20.1) and find a simple yet effective parameterization for Ω, the collision term. We only discuss the simplest model here, the single relaxation time approach (SRT) for a concise introduction, but interested readers are encouraged to look into multiple relaxation time approaches (MRT), which have become mainstay over the past decade or so.

The idea behind the SRT model is that the right-hand side of Eq. (20.1) is not trivially 0 because the system investigated is not in equilibrium. If the rate of relaxation to local equilibrium is exponential (e.g., as in first-order reactions or the nuclear decay), we can frame the collision operator as

$$\Omega(\vec{x}, \vec{v}, t) = -\frac{1}{\tau}\Delta f(\vec{x}, \vec{v}, t), \tag{20.3}$$

where τ is the relaxation time (constant) and Δf is a departure of the probability function from local equilibrium conditions

$$\Delta f(\vec{x}, \vec{v}, t) = f(\vec{x}, \vec{v}, t) - f^0(\vec{x}, \vec{v}, t), \tag{20.4}$$

with $f^0(\vec{x}, \vec{v}, t)$ the local equilibrium probability function. Boltzmann and James Clerk Maxwell showed that the equilibrium probability distribution function in statistical mechanics is the Maxwell–Boltzmann distribution, which is expressed in terms of statistical (upscaled) variables such as averaged velocity field $u(\vec{x}, t)$ and density field $\rho(\vec{x}, t)$. Note that $u \neq v$, the macroscopic momentum here is defined through a statistical average of the particle velocity distribution

$$\rho(\vec{x}, t)\vec{u}(\vec{x}, t) = \int_v \vec{v} f(\vec{x}, \vec{v}, t)\,\mathrm{dv}. \tag{20.5}$$

Similarly, the density field is retrieved by averaging the probability distribution functions

$$\rho(\vec{x}, t) = \int_v f(\vec{x}, \vec{v}, t)\,\mathrm{dv}. \tag{20.6}$$

In the lattice Boltzmann method, the discrete version of the Boltzmann equation is solved on a lattice that provides a structured grid (in space). The lattice also reduces the infinite choice of possible velocity vectors to a finite (small) subset that limits the possible motion of the probability distribution functions from a node to any of its neighbors (see Figure 20.1). The lattice is therefore both a spatial and velocity discretization tool and the number of admissible velocities that the lattice accepts is set by the complexity of the equations that we want to solve and the dimensionality of the problem (one, vs two or three, dimensions).

Discretizing the Boltzmann equation on these lattices and introducing the notation

$$f_i(\vec{x}, t) \equiv f(\vec{x}, \vec{v}_i, t), \tag{20.7}$$

we get

$$f_i(\vec{x} + \vec{v}_i\Delta t, t + \Delta t) - f_i(\vec{x}, t) = -\frac{\Delta t}{\tau}\left(f_i(\vec{x}, t) - f_i^0(\vec{x}, t)\right). \tag{20.8}$$

This equation is quite general for the lattice Boltzmann SRT model and can be used to solve widely different sets of partial differential equations (PDEs). We take the case of a two-dimensional transient diffusion equation to show how the method can be applied to a familiar equation. Most of

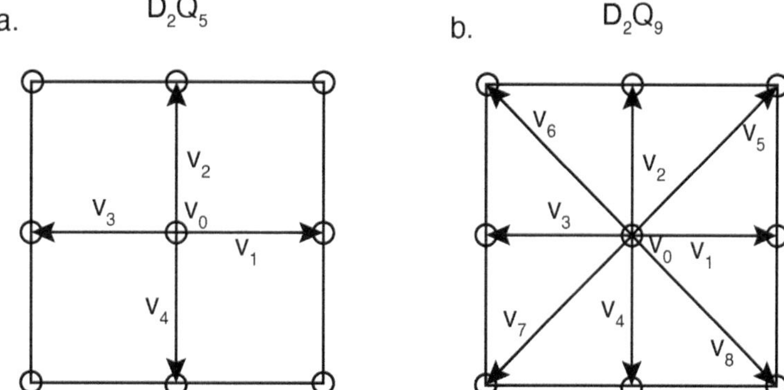

Figure 20.1 Schematic illustration of two common two-dimensional lattices, the D_2Q_5 and D_2Q_9 with respectively 5 and 9 discrete velocities. Lattice nodes are separated by a distance $dx = dy = 1$ in lattice units (by definition). Distributions can only move according to the discrete velocities highlighted by arrows (and including the rest velocity v_0) and always end up at a neighbor lattice node (or stay in place) during a single timestep.

the ideas generalize to more complex sets of PDEs, for example, Navier–Stokes. Let us assume that the scalar field being diffused is $T(\vec{x}, t)$ and that the diffusion process is isotropic for simplicity. Similar to Eq. 20.6, we can relate the scalar field of interest to the distribution

$$T(\vec{x}, t) = \sum_i f_i(\vec{x}, t), \tag{20.9}$$

where the sum is conducted over the different admissible velocities (5 for D_2Q_5). Another important definition is the equilibrium distribution function. Because only one quantity (here related to T) is conserved, it depends solely on T and it is generally defined as

$$f_i^0(\vec{x}, t) = w_i T(\vec{x}, t), \tag{20.10}$$

where w_i are lattice weights that are defined to ensure that the proper symmetry is preserved when the discrete set of velocities is introduced. By construction

$$\sum_i w_i = 1. \tag{20.11}$$

A few important properties arise from these definitions. First, the macroscopic fields are fully encoded by the equilibrium part of the distribution (denoting the nonequilibrium part of the distribution f_i^1)

$$T(\vec{x}, t) = \sum_i f_i = \sum_i w_i T(\vec{x}, t) + \sum_i f_i^1 \quad \rightarrow \quad \sum_i f_i^1 = 0. \tag{20.12}$$

The second property is that, by symmetry of the lattice velocities, we can infer that

$$\sum_i w_i v_{i,\alpha} = 0 \tag{20.13}$$

$$\sum_i w_i v_{i,\alpha} v_{i,\beta} = c_s^2 \delta_{\alpha,\beta}. \tag{20.14}$$

Here c_s is the sound speed of the lattice (1/3 for D_2Q_5) and $\delta_{\alpha,\beta}$ is the Kronecker delta, which is 1 when $\alpha = \beta$ and 0 otherwise and Greek letters indices refer to spatial dimensions (e.g. x,y).

The goal now is to retrieve the diffusion equation at the macroscopic scale from Boltzmann's equation using a specific upscaling procedure known as the Chapman–Enskog expansion. This expansion is based on the definition of a small dimensionless parameter ϵ, which is related to the Knudsen number, that is, the relative measure of the mean free path of particles (i.e., average distance between collisions) to a characteristic macroscopic distance of the problem (e.g., the lengthscale of the domain). The lattice Boltzmann method requires that $\epsilon \ll 1$. The first part of the expansion is to allow for a level of nonequilibrium ($\Delta f_i \neq 0$) but that it remains relatively small (which is required anyway to assume diffusion to be a dominant process). In such a case we can scale the relative importance of the equilibrium and nonequilibrium (f_i^1) parts of the distributions so that

$$f_i(\vec{x}, t) = f_i^0(\vec{x}, t) + \epsilon f_i^1(\vec{x}, t). \tag{20.15}$$

Another important aspect of the Chapman–Enskog expansion is to define the scale for perturbations in space and time, which is done through the definition of the partial derivatives

$$\frac{\partial}{\partial t} = \epsilon^2 \frac{\partial^{(2)}}{\partial t} \tag{20.16}$$

$$\frac{\partial}{\partial x} = \epsilon \frac{\partial^{(1)}}{\partial x}, \tag{20.17}$$

where the superscript (x) does not refer to the order of the partial derivative but rather is there to remind us of the order in ϵ of the derivative. The keen observer will understand this choice as it is natural consequence of a diffusion equation where time scales with the square of the distance.

$$f_i(\vec{x} + \vec{v}_i \Delta t, t + \Delta t) - f_i(\vec{x}, t) = -\epsilon \frac{\Delta t}{\tau} f_i^1(\vec{x}, t). \tag{20.18}$$

The left-hand side can be expanded with a Taylor series to the second order ϵ^2 (omitting location and time when not necessary)

$$\begin{aligned} f_i(\vec{x} + \vec{v}_i \Delta t, t + \Delta t) - f_i(\vec{x}, t) = \epsilon^2 \frac{\partial^{(2)}}{\partial t} f_i^0 + \epsilon \Delta t v_{i,\alpha} \frac{\partial^{(1)}}{\partial x_\alpha} f_i^0 + \\ \epsilon^2 \Delta t v_{i,\alpha} \frac{\partial^{(1)}}{\partial x_\alpha} f_i^1 + \epsilon^2 \frac{\Delta t^2}{2} v_{i,\alpha} v_{i,\beta} \frac{\partial^{(1)}}{\partial x_\alpha} \frac{\partial^{(1)}}{\partial x_\beta} f_i^0 \end{aligned} \tag{20.19}$$

where the summation convection for repeated Greek indices is used

$$v_{i,\alpha} \frac{\partial^{(1)}}{\partial x_\alpha} f_i^1 = v_{i,x} \frac{\partial^{(1)}}{\partial x} f_i^1 + v_{i,y} \frac{\partial^{(1)}}{\partial y} f_i^1. \tag{20.20}$$

At the order ϵ (neglecting all terms in ϵ^2) we get a simple relationship

$$\epsilon \Delta t v_{i,\alpha} \frac{\partial^{(1)}}{\partial x_\alpha} f_i^0 = -\epsilon \frac{\Delta t}{\tau} f_i^1. \tag{20.21}$$

Multiplying both sides by $v_{i,\beta}$ and summing over the velocities yields, after some algebra and using some of the definitions above,

$$\sum_i v_{i,\alpha} f_i^1 = -\tau c_s^2 \frac{\partial^{(1)} T}{\partial x_\alpha}. \tag{20.22}$$

Returning to our expanded Boltzmann equation considering both linear and quadratic terms in ϵ, we have (after summing over the velocities)

$$\epsilon^2 \frac{\partial^{(2)}}{\partial t} \sum_i w_i T + \epsilon \Delta t \frac{\partial^{(1)}}{\partial x_\alpha} \sum_i w_i v_{i,\alpha} T + \epsilon^2 \Delta t \frac{\partial^{(1)}}{\partial x_\alpha} \sum_i v_{i,\alpha} f_i^1$$
$$\epsilon^2 \frac{\Delta t^2}{2} \frac{\partial^{(1)}}{\partial x_\alpha} \frac{\partial^{(1)}}{\partial x_\beta} \sum_i w_i v_{i,\alpha} v_{i,\beta} T = -\epsilon \frac{\Delta t}{\tau} \sum_i f_i^1. \tag{20.23}$$

Using some of the properties associated with the symmetry of the velocities, we get after reorganization

$$\epsilon^2 \frac{\partial^{(2)} T}{\partial t} = \epsilon^2 c_s^2 \Delta t \left(\tau - \frac{\Delta t}{2}\right) \left(\frac{\partial^{(1)^2} T}{\partial x^2} + \frac{\partial^{(1)^2} T}{\partial y^2}\right). \tag{20.24}$$

Using the definition of the derivatives we retrieve the two-dimensional diffusion equation (isotropic)

$$\frac{\partial T}{\partial t} = D \left(\frac{\partial^2 T}{\partial x^2} + \frac{\partial^2 T}{\partial x^2}\right) \tag{20.25}$$

with a diffusion coefficient

$$D = c_s^2 \Delta t \left(\tau - \frac{\Delta t}{2}\right). \tag{20.26}$$

Not surprisingly, the diffusion coefficient value is set by the relaxation time τ. Again the relaxation of the system to equilibrium in the lattice Boltzmann SRT method is exponential with a characteristic time scale defined by τ, hence the transport coefficient D must be set by the relaxation time. Stability of the method requires that $\tau > 1/2$ to ensure that the diffusion coefficient is positive and satisfies a positive entropy production.

This exercise shows a challenge presented by bottom-up approaches that was not present for top-down approaches: it takes work to start from the equations that are modeled (here the discretized Boltzmann equation) to the intended PDE. So a small change in the scheme at the level of the Boltzmann equation needs to be propagated through the expansion to make sure that it does what it was designed for; it may not be as obvious as one thinks.

Note that, although the lattice Boltzmann method is inherently explicit (implicit variations do exist), the diffusion model proposed here is unconditionally stable (with the only requirement that $\tau > 1/2$). Besides this interesting property, the lattice Boltzmann method provides a different viewpoint on the solution of PDEs in that they decompose the dynamics at the particle scale between equilibrium and nonequilibrium, with the former being directly related to the definition of the macroscopic fields of interest and the latter related to fluxes at the continuum scale. This offers valuable advantages when dealing with boundary conditions, especially flux conditions. In general, lattice Boltzmann methods are easy to implement and can be parallelized efficiently.

As a road map to building a model to solve PDEs with the lattice Boltzmann method, it is customary to decompose the time integration into two distinct steps: streaming (which is the left-hand side of Eq. 20.8) and collision. The interested reader is referred to the detailed textbooks of [4, 23, 24] for more information about the method and other applications.

Part V

Acknowledgments

When Oxford University Press first approached me about writing a textbook on computational methods, about a decade ago, my hesitation came mostly from the fact that I held a junior (pre-tenure) faculty position (then at Georgia Tech) and that the effort of writing a textbook from scratch seemed daunting, especially at a time when one is expected to build a group and demonstrate productivity. I was naive enough to accept the challenge. While it took much longer than I expected, this experience has been refreshing and exciting. As I taught more classes on the topics presented in this book, I was able to expand several sections and to seek new didactic ways to explain some of the challenging concepts that you find in this book. Retreating (even for a couple of hours) to write a subsection of this book was a quiet and meditative experience in an otherwise busy and overly stimulated work routine. I hope each reader finds this book to be a good companion as they learn a difficult but fascinating tool (numerical modeling) that can enhance their learning and possibly research experiences. When writing this book my goal was to develop a self-contained reference to initiate readers to mathematical modeling in the field of physical sciences, including the mathematical foundations, as well as to introduce some of the physics concepts related to the equations that we cover. A secondary goal was to provide a somehow rigorous mathematical reference that provides good habits to beginners and reminders to more advanced readers.

As I eluded, this labor of love took some time—far beyond the expected deadline of Oxford University Press. I acknowledge the patience of the different editors I have worked with, starting with Sonke Adlung early on and Giulia Lipparini over the past few years. Their patience and constant encouragement are truly appreciated and kept me in good spirits during the writing of this book. I also thank students who read some early sections of the book and provided comments or found typos (I am sure there are more to be found!) and who improved the manuscript as it progressed. These include graduate students working with me over the past years, Hamid Karani, Majid Rasht-behesht, Laura Lark, Darien Florez, Erica Nathan, and Nina Gilkyson, as well as students who took my intro to numerical modeling class at Georgia Tech and Brown University over the years. I also thank Gregorio Posada Pardo for going through the manuscript and compiling the index as well as finding typos.

Finally, I thank my wife Olga and our daughters Bénédicte and Noémie for their patience and support. Another big thank you to Olga for her artistic inputs, and a few figures throughout the text.

Bibliography

1 Francis Albarède. *Introduction to Geochemical Modeling*. Cambridge University Press, 1996.

2 Robert S. Anderson, and Suzanne P. Anderson. *Geomorphology: The Mechanics and Chemistry of Landscapes*. Cambridge University Press, 2010.

3 Sumit Chakraborty. Diffusion in Solid Silicates: A Tool to Track Timescales of Processes Comes of Age. *Annual Review of Earth and Planetary Sciences* 36(1):153–90; 2008.

4 Bastien Chopard, and Michel Droz. *Cellular Automata of Modelling of Physical Systems*. Cambridge University Press, 1998.

5 Fidel Costa, and Daniel Morgan. "Time Constraints from Chemical Equilibration in Magmatic Crystals." *Timescales of Magmatic Processes: From Core to Atmosphere*, pages 125–59, Anthony Dosseto, James A. Van-Orman, and Simon P. Turner. Wiley, 2010.

6 Fidel Costa, Thomas Shea, and Teresa Ubide. "Diffusion Chronometry and the Timescales of Magmatic Processes." *Nature Reviews Earth & Environment* 1(4): 201–14; 2020.

7 John Crank. *The Mathematics of Diffusion*. Oxford University Press, 1979.

8 William Morris Davis. "The Convex Profile of Badland Divides." *Science* 508: 245; 1892.

9 Christina L. de La Rocha, and Donald J. DePaolo. "Isotopic Evidence for Variations in the Marine Calcium Cycle over the Cenozoic." *Science* 289(5482): 1176–8; 2000.

10 C. Bouvet de Maisonneuve, Fidel Costa, Christian Huber, Pierre Vonlanthen, Olivier Bachmann, and Michael A. Dungan. "How do Olivines Record Magmatic Events? Insights from Major and Trace Element Zoning." *Contributions to Mineralogy and Petrology* 171: 1–20; 2016.

11 Nelson F. Fernandes and William E. Dietrich. "Hillslope Evolution by Diffusive Processes: The Timescale for Equilibrium Adjustments." *Water Resources Research* 33(6): 1307–1318; 1997.

12 Grove Karl Gilbert. "The Convexity of Hilltops." *Journal of Geology* 27: 344–50; 1909.

13 Alan D. Howard. "A Detachment-limited Model of Drainage Basin Evolution." *Water Resources Research* 30(7): 2261–85; 1994.

14 Herbert E. Huppert and Andrew W. Woods. "The Role of Volatiles in Magma Chamber Dynamics." *Nature* 420(6915): 493–5; 2002.

15 A. C. Lasaga. "The Kinetic Treatment of Geochemical Cycles." *Geochimica et Cosmochimica Acta* 44(6): 815–28; 1980.

16 Peter D. Lax, and Robert D. Richtmyer. "Survey of the Stability of Linear Finite Difference Equations." In *Selected Papers Volume I*, pages 125–51, edited by Peter D. Lax, Peter Sarnak, and Andrew J. Majda. Springer, 2005.

17 Randall J. LeVeque. *Finite Volume Methods for Hyperbolic Problems*. Cambridge Texts in Applied Mathematics. Cambridge University Press, 2002.

18 John Pelletier. *Quantitative Modeling of Earth Surface Processes*. Cambridge University Press, 2008.

19 J. Taylor Perron. "Numerical Methods for Nonlinear Hillslope Transport Laws." *Journal of Geophysical Research: Earth Surface* 116(F2); 2011.

20 I. Prigogine and R. Lefever. "Symmetry Breaking Instabilities in Dissipative Systems. II." *The Journal of Chemical Physics* 48(4): 1695–1700; 1968.

21 Joshua J. Roering, James W. Kirchner, and William E. Dietrich. "Evidence for Nonlinear, Diffusive Sediment Transport on Hillslopes and Implications for Landscape Morphology." *Water Resources Research* 35(3): 853–70; 1999.

22 Thomas Shea, Fidel Costa, Daniel Krimer, and Julia Eve Hammer. "Accuracy of Timescales Retrieved from Diffusion Modeling in Olivine: A 3D Perspective." *American Mineralogist* 100(10): 2026–42; 2015.

23 Sauro Succi. *The Lattice Boltzmann Equation: For Fluid Dynamics and Beyond.* Oxford University Press, 2001.

24 M. C. Sukop and D. T. Thorne Jr. *Lattice Boltzmann Modeling: An Introduction for Geoscientists and Engineers.* Springer, 2006.

25 Loup Verlet. "Computer 'Experiments' on Classical Fluids. i. Thermodynamical Properties of Lennard-Jones Molecules." *Physical Review* 159(1): 98; 1967.

26 Andrew J. Watson and James E. Lovelock. "Biological Homeostasis of the Global Environment: The Parable of Daisyworld." *Tellus B: Chemical and Physical Meteorology* 35(4): 284–9; 1983.

27 Youxue Zhang. "Diffusion in Minerals and Melts: Theoretical Background." *Reviews in Mineralogy and Geochemistry* 72(1): 5–59; 2010.

28 Tarek I Zohdi. *A finite element primer for beginners: the basics.* Springer, 2015.

29 E. C. Zachmanoglou and Dale W. Thoe. *Introduction to Partial Differential Equations with Applications.* Courier Corporation, 1986. ISBN 0-486-65251-3

Index